Interpretation of Geometric Dimensioning and Tolerancing

Interpretation

of

Geometric
Dimensioning

and

Tolerancing

Daniel E. Puncochar

Industrial Press Inc.
200 Madison Avenue
New York, New York 10016-4078

Library of Congress Cataloging-in-Publication Data

Puncochar, Daniel E.
    Interpretation of geometric dimensioning and tolerancing / Daniel
E. Puncochar.
      p. cm.
    ISBN 0-8311-3010-5
    1. Tolerance (Engineering) 2. Dimensions. I. Title.
T357.P96 1990
604.2'43—dc20

89-39412
CIP

Industrial Press Inc., 200 Madison Avenue, New York, New York 10016-4078

First Printing

Interpretation of Geometric Dimensioning and Tolerancing

Composition, printing and binding by Edwards Brothers, Inc., Ann Arbor, Michigan

9 8 7 6 5 4 3 2

Preface

Geometric dimensioning and tolerancing (G.D.T.) is a method for stating and interpreting design requirements. G.D.T. is an international system of symbolic language, and is simply another tool available to make engineering drawings better means of communication from design through manufacturing and inspection. Some of G.D.T.'s advantages are uniformity in design practice, fewer misinterpretations, ensured interchangeability, and maximum tolerance allocation. Also, with G.D.T., design requirements are specified explicitly and the latest gaging techniques are accommodated. These advantages contribute to higher production yields with less rework or scrap.

To help the reader understand G.D.T., *Interpretation of Geometric Dimensioning and Tolerancing* begins with basic principles and builds on these principles with applications-oriented concepts. Complex material is presented in a "building-block" approach, with examples that illustrate each concept. End-of-the-chapter examples further reinforce the explanations given in each section. It is assumed that each reader has an adequate knowledge of blueprint reading. Since G.D.T. applies regardless of the measurement system used, some drawing examples are dimensioned and toleranced in inches and some in millimeters.

Interpretation of Geometric Dimensioning and Tolerancing completely covers the material in ANSI Y14.5M-1982, but does not prescribe design practices, state design requirements, specify inspection techniques, or specify any other engineering practice. However, it is sometimes necessary to state how something is specified or inspected so that a concept can be discussed adequately. In addition, the drawings in this text are not complete production drawings, but only present the concepts currently under discussion.

It is hoped that *Interpretation of Geometric Dimensioning and Tolerancing* will assist the reader in becoming conversant in the techniques of G.D.T. given in the ANSI standard—techniques that can be integrated smoothly into CAD/CAM and modern inspection systems.

Acknowledgments

I am grateful to the management team at Freightliner Corporation for the opportunity to work with G.D.T. in more depth. I am especially grateful to Dr. Floyd Hunsaker for all of his support and encouragement. Without the opportunity and support provided by the Freightliner management, the writing of this book would have been formidable.

The efforts and support of two special friends must be recognized. Dr. Sam Stern, of Oregon State University, and Greg Hutchins, Professional Engineer, provided me with the moral support, proper guidance, and motivation throughout the process. Their encouragement, assistance, writing experience, and general support were invaluable.

This book would only be an organized group of words without illustrations to assist in making particular points. The illustrations were prepared by Donald L. Knotts, Jr., Design Engineer.

Finally, I am deeply indebted to my wife Margaret, and our family for their support and patience during this writing—especially to Margaret for the typing and proofreading of the manuscript. All of the family assisted in some way by carrying on various tasks and chores without my assistance.

Contents

Contents _____ *ix*

Interpretation of Geometric Dimensioning and Tolerancing

Introduction

History

During the early period of manufacturing there seldom were any drawings. A person had an idea for something that was needed for industry, farming, or mining, and made it. Usually there was only one item, and when repair was needed, someone repaired or replaced the needed part right on the job.

Over time, complexities in manufacturing increased, and there was an increased need for drawings of parts and their assemblies. With drawings came tolerances—parts were permitted some variation rather than being fitted to only one assembly. This tolerance was specified as a plus/minus tolerance.

This plus/minus or coordinate system worked quite well and still does for many applications. But, today, the need for interchangeability of parts is very important, with parts and assemblies manufactured around the world being brought together at an assembly plant that makes automobiles, tractors, airplanes, televisions, projectors, space craft, and so on. These items also need replacement parts that assemble readily without a lot of rework.

As the demand for parts manufactured around the world grew, so did the need for accuracy. Accuracy became more critical because of competition for parts and assemblies. The idea of positional tolerancing, which provided a means for locating round features within a round tolerance zone rather than the traditional square tolerance zone, was introduced. The idea caught on and was adopted by the military. It became part of the military standards and later was a Unified American Standard Association standard, ASA Y14.5. This standard was released in 1956 and was accepted by the military. Later ASA became the American National Standards Institute (ANSI). ANSI later published a complete system of symbology for geometric form and positional tolerances, "Dimensioning and Tolerancing."

In 1983, the current standard, ANSI Y14.5M-1982, was released. This standard clarifies some of the old practices and moves a little closer to the practices of the International Organization of Standards (ISO). ISO is primarily a European standard. Today, geometric dimensioning and tolerancing (G.D.T.) is used by the majority of larger manufacturing companies in the United States and the world.

The Importance of Standards

We live and work in accordance with standards all of the time. Almost every aspect of life, education, or business operates according to some standard. Some of the standards are specified and controlled locally, while others are national and still others are international.

WHAT IS A STANDARD?

A standard is a model or rule with which other similar things are to be made to or compared; G.D.T. is an example of such a standard. The G.D.T. symbols are the model that is provided internationally. This system was created to improve communications, control, and productivity in manufacturing. Standards are critical to all of us, and they will become increasingly important as our technologies continue to develop.

CHANGE

Change is one of the most constant things we have as a society. As a result, standards change also, especially those associated with technology. G.D.T. is an example of a widely used standard that must be updated constantly to be useful to industry. Over time, needed improvements are identified, discussed, and evaluated. When a sufficient amount of improvements are agreed upon, standards are changed. The most recent update to the G.D.T. standard was adopted in 1982.

UNIVERSALITY

G.D.T. is a standard throughout the world. In the United States the standard is maintained by the American National Standards Institute. In the European countries and other parts of the world, the standard is maintained by the International Organization for Standards. These two standards are not identical, but with each revision the standards become more similar. As our international companies continue to exchange drawings among themselves, these standards will become increasingly important.

Why G.D.T.?

G.D.T. adds clarity and contributes its many advantages to our coordinate system of dimensioning. The old system (coordinate) was lacking in many respects. A part of the designer's intent was always left to interpretation by the craftsman. Probably the most significant difference between the two systems is the location of round features; for example, the coordinate system had a square tolerance zone, which allowed good parts to be rejected. In our world of high technology and transfer of parts around the world, we cannot tolerate the misinterpretation that is possible with the coordinate system.

G.D.T. IS NOT A REPLACEMENT

The coordinate system is not being replaced entirely with G.D.T. G.D.T. is specified to enhance the coordinate system as required per design. When the advantages of G.D.T. can be utilized, they are simultaneously specified. G.D.T., a system of symbols, provides a means of completely specifying uniformity and describing the designer's intent. These symbols eliminate most of the drawing misinterpretation by not having notes in drawing margins and by having complete descriptions of features and design requirements.

COMPLETE SPECIFICATION

A complete specification of design requirements is possible with symbols. These symbols also allow the designer to specify maximum tolerances for parts that must assemble with other parts. These maximum tolerances also ensure the interchangeability of parts. The use of symbols for complete specification is becoming increasingly important with the growing interrelated ownerships of companies around the world. G.D.T. is fast becoming an international "common symbolic language" controlled by standards. Today, the majority of U.S. manufacturing companies are applying G.D.T. to their drawings.

ADVANTAGES

There are many reasons for specifying geometric tolerancing wherever design integrity must be controlled and communicated completely to others. Two key principles for applying G.D.T. are the function and the relationship of parts in an assembly. Probably the most advantageous part of G.D.T. is the method of specifying feature location. In the past, features were located with the coordinate system.

The coordinate system is a method of tolerancing that uses a plus/minus tolerance. Plus and minus tolerances are specified for lengths, widths, diameters, shapes, and locations. An illustration of how drawings may be dimensioned and toleranced with the coordinate system is shown in Fig. 1-1. This method of tolerancing permits the length and diameters to vary by a plus/minus value. It also allows the maker of the part to put the center hole wherever he or she desires. By looking at the drawing we can only assume that the hole is centered. This is an example of a drawing being left open to misinterpretation by anyone reading it.

A similar situation exists where hole or pin locations are specified with the coordinate system. An example of how holes are specified is shown in Fig. 1-2. The tolerance as specified establishes a square tolerance zone based on the plus/minus five thousandths of an inch tolerance in the x and y directions. There is no consideration for the actual feature size.

The tolerance zone is ten thousandths of an inch on each side of a square regardless of the actual feature size. An illustration of how the tolerance zone appears is shown in Fig. 1-3.

The axis of the hole or pin must be positioned in

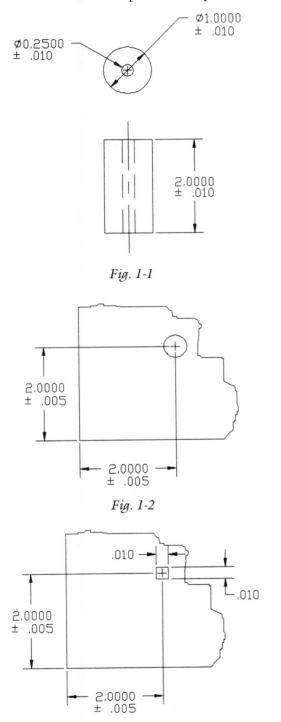

Fig. 1-1

Fig. 1-2

Fig. 1-3

that square zone in order for the feature to be located properly. The feature may lean or slant an uncontrolled amount as long as the axis stays within the square zone. The designer only assumes that the feature will be produced nearly perpendicular to the material it is put into. If the axis were in an extreme corner of this zone, the feature location is still acceptable; that radial measurement is .007 in., as shown in Fig. 1-4.

The only place there is a .007 in. measurement is from the center to any corner. That .007 in. should be usable all around the desired true position, as illustrated in the example in Fig. 1-5. G.D.T. provides a method of specifying a tolerance zone that takes the shape of the feature if so desired by the designer. G.D.T. also allows consideration of feature size for calculating total tolerance. This concept is presented in later chapters.

Other advantages of G.D.T. include international uniformity in describing designer intent. The symbolic method of specifying design intent eliminates most misinterpretation of drawing notes. In the past drawings usually contained in the margin a list of notes that were intended to explain certain requirements. These notes were all subject to misinterpretation. With the available symbols, designers can more readily specify complete design requirements. The proper application of geometric tolerances ensures the interchange-

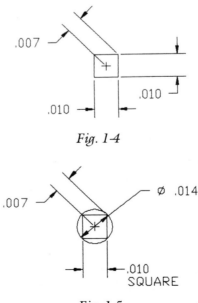

Fig. 1-4

Fig. 1-5

ability of parts. G.D.T. is very rapidly becoming a common language throughout industry internationally.

Chapter 1 Evaluation

1. Blueprints are the primary _____ tool between designers and manufacturing.

2. G.D.T. is a system made up of _____ primarily.

3. The G.D.T. standard is one of the standards maintained by the _____.

4. G.D.T. adds to the coordinate dimensioning system when specific _____ is required.

5. G.D.T. does not _____ the coordinate dimensioning and tolerancing system.

6. G.D.T. can best be described as a _____ dimensioning and tolerancing system.

7. G.D.T. is used to control the _____ of a part feature and its relationship to other features.

8. G.D.T. is also used to control feature _____ and _____.

9. The key principles of G.D.T. are _____ and _____.

10. Two advantages of the G.D.T. system are maximum _____ and ensured _____ of mating parts.

11. The total amount that a part size may vary is a size _____.

12. A common method used to specify a tolerance for the nominal size of a feature is _____ values.

Symbols and Abbreviations

Introduction

Geometric dimensioning and tolerancing (G.D.T.) is a language of symbols. This chapter will introduce these symbols and give an application for each of them. Throughout the remainder of this book you will see these symbols applied to various drawings in combination with other symbols. The symbols presented here are the ones used to specify geometric characteristics and dimensional requirements on industrial drawings, which are in accordance with ANSI Y14.5M-1982.

One of the advantages of G.D.T. is that it is an international language of symbols that generally eliminates the need for drawing notes. (See Appendix B for definitions of the symbols.) However, in limited situations it may be necessary for the designer to add a short note to aid in conveying the design requirements. In Figs. 2-1 and 2-2 are two examples of when a note may be specified.

Symbols

DIAMETER

The term diameter is very commonly used with many aspects of our lives. Most of us are familiar with the abbreviations D or DIA. Now with G.D.T. we also have a symbol for diameter, it is a circle with a slash through it, $\emptyset$. The diameter symbol is used to describe cylindrical features and tolerance zones.

This symbol always precedes the size or tolerance specification. In Fig. 2-3 is an example of how this symbol appears in application. **Note:** The absence of this symbol before either the feature size or tolerance indicates a noncylindrical feature or tolerance; see Fig. 2-10.

BASIC

Basic is a term used to describe the theoretically exact size, shape, or location of a feature or datum. Basic dimensions _do not_ have a tolerance. Tolerances for basic dimensions are specified in feature control frames, or tooling tolerances apply under other conditions. Basic can be abbreviated as BSC, or, most recently, the use of the basic symbol is used; the symbol for basic is a rectangle around the dimension.

Basic is used to describe the theoretically exact shape or profile of surfaces either regular or irregular. Most frequently, basic is used to specify the exact desired position of features. Then, each feature is given a tolerance that allows some variation depending on the design requirement. Figure 2-3 shows an example of how basic may be specified.

MAXIMUM MATERIAL CONDITION

The next symbols and abbreviations to be introduced are those for the modifiers. The modifiers are maximum material condition, least material condition, and regardless of feature size. The first two may modify a specified tolerance, as their name indicates. In the third one, as its name indicates, the specified tolerance is used regardless of the feature size. The first modifier to be discussed in this chapter is maximum material condition.

The symbol for maximum material condition is a capital "M" in a circle, Ⓜ. Maximum material condition may also be abbreviated MMC. Most frequently, the symbol is specified on the drawing, and the abbreviation is used during discussions about tolerancing.

The term maximum material condition is used to describe the maximum condition of a feature of size.

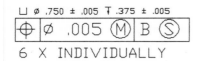

6 X INDIVIDUALLY

Fig. 2-1

For example, a hole is a feature of size that is permitted to vary in size within the limits of a plus/minus tolerance. For holes or any internal feature, MMC is the smallest size for that feature. In other words, the maximum material remains in the piece the hole was put in. An example of a hole size specification that permits an MMC of ∅.518 is shown in Fig. 2-3. At .518 diameter the positional tolerance is ∅.026.

Maximum material condition is also applicable to external features of size, such as pins, tabs, and splines. With these features MMC is equal to the largest size permitted by the size specification. An example of a pin size that permits an MMC of ∅.512 is shown in Fig. 2-4.

The application of this modifier would include designs where maximum tolerances are needed, such as in the assembly of parts, where maximum tolerances are desirable to permit the interchangeability of parts. Maximum tolerances are achieved as the feature departs from MMC. A minimum tolerance is stated and then a modifier symbol is added following the tolerance when applicable to permit an increase in tolerance equal to the departure from MMC. This concept will be covered thoroughly in Chapters 6 and 8.

LEAST MATERIAL CONDITION

The next modifier is least material condition. The symbol is a capital "L" in a circle, Ⓛ. As with maximum material condition, the symbol is used on drawings and an abbreviation LMC is used during tolerance discussions.

The term least material condition is used to describe the opposite condition as maximum material condition. This modifier is specified for features of size to describe the least material condition. For ex-

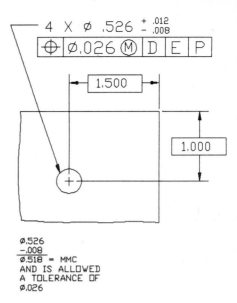

Fig. 2-3

ample, using the same hole call out as in Fig. 2-3, the LMC is equal to ∅.538 (Fig. 2-5). Here the least amount of material remains in the plate after the hole is put in it.

Least material condition also applies to external features of size. For external features subject to size variation, LMC means these features contain the least material within the specified size limits. In Fig. 2-6 an example of a pin at least material condition is shown.

The application of this modifier would include designs where a minimum material thickness must be maintained, such as holes near material edges or bosses with holes bored in them where the wall thickness is critical. This modifier provides the same advantages as MMC but in the opposite direction.

BETWEEN Y & Z

Fig. 2-2

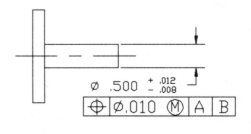

Fig. 2-4

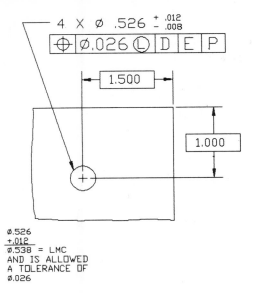

Fig. 2-5

REGARDLESS OF FEATURE SIZE

The final modifier to be covered is regardless of feature size. The symbol is a capital "S" in a circle, Ⓢ. This modifier may be abbreviated RFS for discussional purposes; the symbol is specified on drawings.

Regardless of feature size is considered a modifier even though it does not permit any modification of stated drawing tolerances. As the name indicates, regardless of the feature's size the stated tolerance applies.

This modifier is a very restrictive requirement. In general, this modifier is seldom specified, because it does restrict the tolerance to the stated tolerance only.

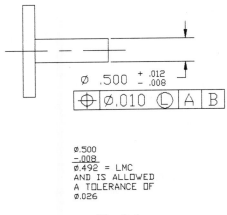

Fig. 2-6

This modifier is specified with tolerances for features of size as the other two modifiers are specified; this one just limits the tolerance to that specified by the designer. An example of a regardless of feature size specification is shown in Fig. 2-7.

Regardless of feature size also applies to external features of size. External features here might be splines or gears and similar types of features that are subject to size variation. With these types of features, designs usually cannot tolerate play or additional tolerance between mating parts. In Fig. 2-8, an example is shown of how RFS may be appled to an external feature.

The application of this modifier would include designs where tolerance allowances are critical. This modifier may be specified for parts that form an assembly with limited shift between parts. Typical applications for restricted tolerances are gears, splines, and press fits.

FULL INDICATOR MOVEMENT

The term full indicator movement is a new term for older terminology. This new term replaces total indicator movement (TIM) and total indicator reading (TIR). There is no symbol for full indicator movement; it is abbreviated FIM. This abbreviation does not appear on drawings. FIM is understood or implied for certain form controls, which control cylindrical features. FIM will be discussed further in

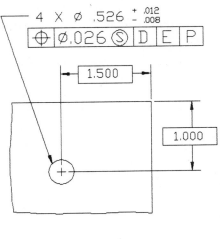

Fig. 2-7

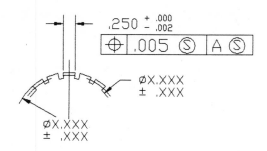

Fig. 2-8

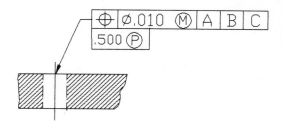

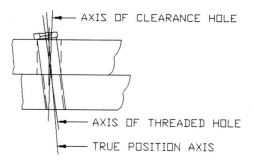

Chapter 6. An example of how FIM is used on a drawing is shown in Fig. 2-9. Figure 2-9 means that the stated tolerances of .002 and .005 are FIM. For example, FIM is the complete movement of a needle on a dial indicator. To measure the variation of the .500 and 1.000 inch diameters, a dial indicator would be rested on each surface and then that surface rotated 360 degrees. During this complete rotation the dial indicator reading must not exceed the stated tolerance.

Fig. 2-10

PROJECTED TOLERANCE ZONE

Projected tolerance zone is not abbreviated; the symbol for it is a capital "P" in a circle, ⓟ. The projected tolerance zone is a method of describing a tolerance, usually for a fixed fastener application, to prevent interference between mating parts.

The specification of a projected tolerance zone is to control the axis of a hole. As shown in Fig. 2-10, the axis of the hole must be controlled within a specified tolerance above the part the hole is in. The tolerance zone is projected to a height equal to the

thickness of the mating part. The purpose of this control is to prevent the hole for the fixed fastener from interfering with the mating part.

DIMENSION ORIGIN

The term dimension origin is not abbreviated, and its symbol is ⊕→. This symbol is used to identify the surface or feature where the dimension originates.

Some designs are complex, thus difficult to determine where dimensions are to begin. In these situations the designer specifies where the dimension is to originate. Usually the part will be different if made by starting dimensions from a different surface or feature than intended. The example in Fig. 2-13 illustrates such a part.

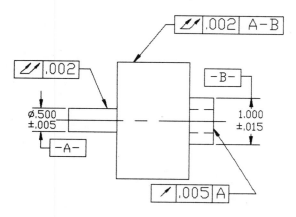

Fig. 2-9

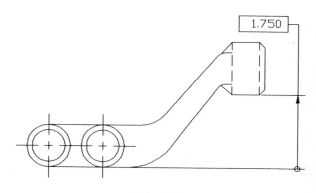

Fig. 2-11

ALL-AROUND

All-around is a term that is not abbreviated. The symbol is ⟲. This symbol is specified along with profile specifications. It means that the profile tolerance applies all around the controlled feature and replaces the words "all-around."

This symbol should only be applied when a uniform profile tolerance is required all around the feature; also, the tolerance is the same around the feature. Figure 2-12 shows an example of how the uniform tolerance applies.

RADIUS

Radius like diameter is another older term that was used on industrial drawings. Radius is not abbreviated; the letter "R" is used as the symbol.

Radius is applied to designs that require the removal of edges or to rounded features. The letter "R" precedes the radial dimension. When a radius is dimensioned in a view that does not show the true shape of the radius, "TRUE R" precedes the radial dimension. See the application of radius in Fig. 2-12. The actual radius must be within the limits of size and must have perfect form with no flats or reversals.

REFERENCE DIMENSION

Another term that is not new to drawings is reference dimension. This term was usually abbreviated to REF and is now represented with a value in parenthesis, ( ).

Reference dimensions are specified for a relationship between features in a flat pattern application. It is not used to define parts. Reference dimensions may be the sum of several dimensions, the size or thick-

ness of material, and specify travel of moving parts, etc. See the illustration in Fig. 2-13.

SPHERICAL DIAMETER

This term was introduced to G.D.T. with the most recent update. This term is abbreviated as SD, and the symbol is S∅.

Spherical diameter is specified for round features. It specifies the diameter of these features. The abbreviation or symbol is specified either before or following the round feature size (Fig. 2-14).

SPHERICAL RADIUS

This term was introduced with the latest update to G.D.T. This term may be abbreviated SR; there is no symbol.

Spherical radius is applied to round features. The abbreviation is specified before the radial value of the feature. See application in Fig. 2-15.

ARC LENGTH

Arc length is a term used to describe the length of a curved surface. The symbol is ⌢, and there is no abbreviation. This symbol is placed over the dimension.

The arc length symbol is specified when it is required to measure along the actual part surface. When this symbol is specified, a linear measurement across the arc is not permitted. In the past, terms like "TRUE" or "TRUE ON SURFACE" were used. See example in Fig. 2-16.

CHAIN LINE

Chain line is a term used to describe or identify a specific area, surface, or portion of a part for special

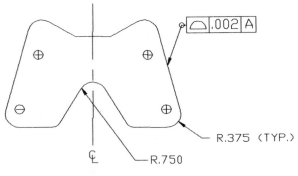

Fig. 2-12

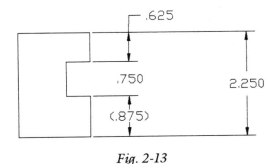

Fig. 2-13

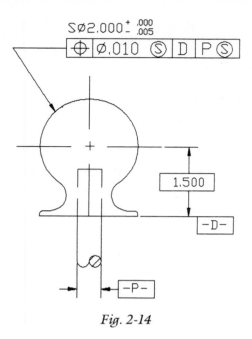

Fig. 2-14

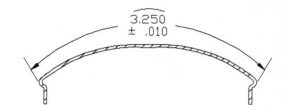

Fig. 2-16

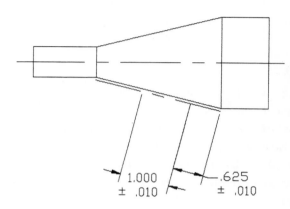

Fig. 2-17

treatment. The term has a symbol, which is ———, and there is no abbreviation.

The chain line symbol is applied when the designer requires only a limited portion of surfaces or areas to be treated differently from the rest of the part. See Fig. 2-17.

CONICAL TAPER

The term conical taper was introduced with the 1982 update to the standard. There is no abbreviation; the symbol is ▷.

There are three methods of specifying a conical taper. First it may be specified with basic dimensions for the diameters and the taper (Fig. 2-18). There may be a size and profile tolerance combined with a profile of the surface (Fig. 2-19). Both diameters along with the length may be toleranced (Fig. 2-20).

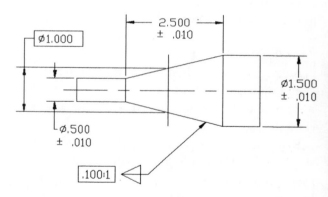

Fig. 2-18

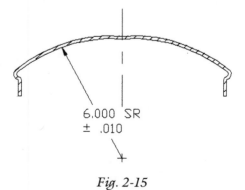

Fig. 2-15

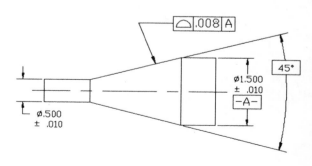

Fig. 2-19

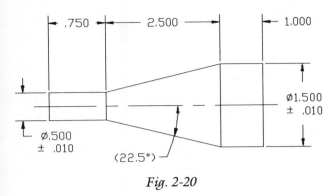

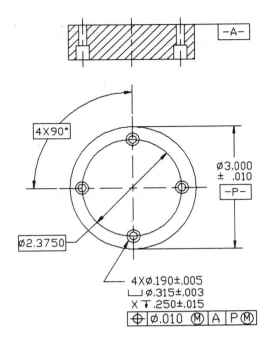

Fig. 2-20

SLOPE

The term slope was introduced to G.D.T. in the 1982 update. There is no abbreviation, and the symbol is ◁.

Slope is primarily specified to control flat tapers. Slope is not specified as degrees, but as a ratio of height differences from one end of the flat taper to the other end. An example of how slope is specified is shown in Fig. 2-21.

COUNTERBORE/SPOTFACE

The terms counterbore and spotface are not new to industrial drawings. The symbol is new, however; it is ⌴. The terms may also be abbreviated as in the past—C'Bore and SF. The symbol is the preferred method to use with G.D.T. application.

Counterbore or spotface may be specified for features that require a recessed or flat mounting surface for fasteners. A typical callout for a counterbore or spotface using the symbol is shown in Fig. 2-22.

COUNTERSINK

Countersink, like the previous terms, is not new to industrial drawings. Countersink may either be abbreviated as C'Sink, or specified with the symbol ∨.

Fig. 2-22

Countersinks are specified for features that require a smooth flat surface. Countersinks are typically specified as shown in Fig. 2-23.

DEPTH/DEEP

The terms depth or deep are not new to industrial drawings. These are words that were written out on

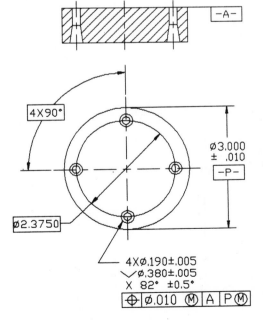

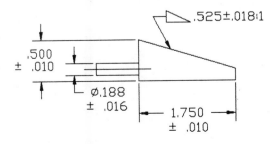

Fig. 2-21

Fig. 2-23

drawings. Now there is a symbol for them, ⊤, or the abbreviation DP is used.

Depth or deep is specified for features that do not pass completely through a part. The symbol may be specified as in Fig. 2-22 for counterbores or it may be specified for blind holes as shown in Fig. 2-24.

DIMENSION NOT TO SCALE

At times it is desirable or necessary to specify dimensions not to scale. The previous practice was to underline the dimension with a wavy line; the current practice is to use a heavy straight line under the dimension.

The symbol for dimensions not to scale is specified when dimensions are intentionally drawn out of scale. Today, with computer-aided design (CAD), there is seldom a situation where out-of-scale dimensioning is used. An example of out of scale dimensioning is illustrated in Fig. 2-25.

TIMES/PLACES

These terms are not really new to drawings, they just have a little different application now. There is no abbreviation, the symbol is an ×.

The designer may specify the number of times or places for patterns of holes, for example, as shown in Fig. 2-26. The same symbol is specified, for example, when a slot size is specified. See the example in Fig. 2-27.

Summary

This completes your introduction to the majority of the symbols and abbreviations associated with geo-

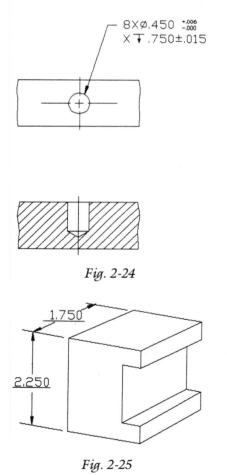

Fig. 2-24

Fig. 2-25

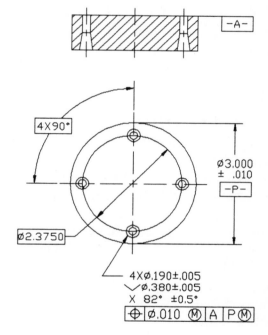

Fig. 2-26

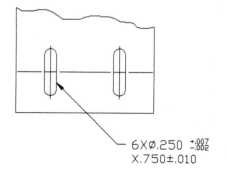

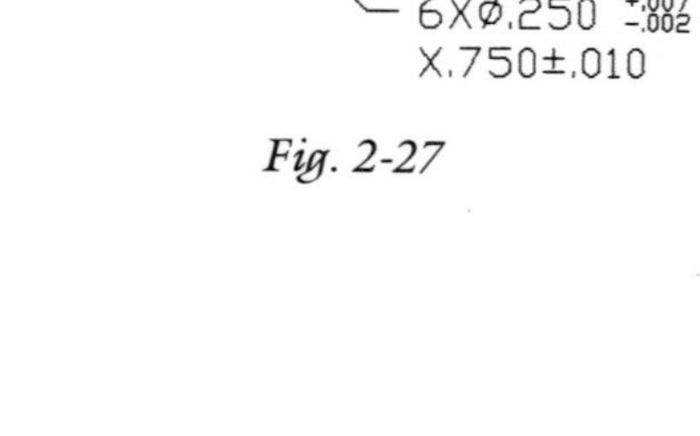

Fig. 2-27

metric dimensioning and tolerancing. There are other symbols that will be introduced in other chapters. Those chapters cover datums, form/orientation controls, and tolerances of location.

The symbols and abbreviations introduced in this chapter will be applied to other drawings and examples throughout this book. They were introduced in this chapter as a basis for the other material.

Chapter 2 Evaluation

Match the abbreviations and symbols in the right column with the phrases in the left-hand column. The abbreviations and symbols may be used more than once and some of them not at all.

____	1. The abbreviation for full indicator movement.	1. ×
____	2. The condition of an internal feature when it measures the largest size within design limits or weighs the least.	2. Ⓛ
____	3. The symbol specified to indicate diametrical or circular features.	3. **SR**
____	4. Regardless of feature size within the design limits of the tolerance.	4. **SØ**
____	5. Dimensions that are for reference only.	5. Ø
____	6. Dimensions that do not have tolerances specified with them.	6. [50]
____	7. The symbol to indicate diametrical tolerance zones.	7. **R**
____	8. The symbol specified to indicate the number of times or places, or by.	8. ↧
____	9. A curved surface that is to be measured along the curve is specified with which symbol?	9. ⌴
____	10. Dimensions that are not to scale are identified with which symbol?	10. Ⓢ
____	11. Parts or features requiring a rounded edge or corners are identified with which symbol?	11. ⌳
____	12. Features that are spherical shaped are identified with which symbol?	12. ⦶→
____	13. The symbol used to indicate the depth of a feature.	13. ↙⊖
____	14. The condition of an external feature when it measures the largest or weighs the most.	14. ⌲
____	15. A slope or flat taper.	15. ———
____	16. Symbol for conical tapers.	16. Ⓟ
____	17. To indicate a countersink.	17. FIM
____	18. Indicates the origin of a dimension.	18. TIR
____	19. Indicates an area for special treatment or processing.	19. **(50)**
____	20. To control the perpendicularity of a fastener to a given height.	20. —
____	21. A counterbore or spotface is specified with which symbol?	21. Ⓜ
____	22. Dimensions that specify theoretically exact size, shape, or location of features.	22. $\overset{\frown}{\mathbf{105}}$
____	23. The total movement of a dial indicator needle.	23. ⌵

Datums

Introduction

Datums provide the framework from which part features are manufactured, and they are critical for all industrial drawings that are created for today's production. Parts assemblies and processes are becoming more critical for profitability. The methods and practices that were acceptable in the past are no longer acceptable with today's technological advancements. Parts produced today must be 100% changeable with their counterparts. The designer can ensure complete fit, function, and interchangeability by clearly specifying the design intent. Datums provide the clarity required for proper feature orientation, which originates from datum planes. Datums are specified on drawings so that repetitive measurements can be made from design through the production and inspection processes. Datums are nothing more nor less than physical features of parts to make repeatable measurements from. These features are considered theoretically exact. Datums are the locating points or surfaces used during the manufacturing process.

Datum Identification

Datum identification is required with ANSI Y14.5 M-1982. Previously, implied datums were permitted. Too often with implied datums, the same surfaces, edges, or features were not used throughout the design, manufacturing, and inspection processes. This can be illustrated with a simple part. A drawing specification for a simple flat part with three holes in it is shown in Fig. 3-1. This part may not be oriented the same through manufacturing and inspection, causing it to be rejected. To illustrate, consider that the part was manufactured using the left-hand end as the surface to butt up with a fence, and the lower edge against a stop. The holes were put in the part while it was set up as illustrated in Fig. 3-2. Then the part was set up for inspection using the longest surface to make contact first. The part would be inspected as shown in Fig. 3-3. When inspected by being set up this way, the part was rejected because the lower-left-hand hole was out of design specification. Therefore, datums must be specified so the drawing is interpreted the same by all who read it.

WHAT IS A DATUM?

A datum is a theoretically exact line, surface, point, area, or axis that is used as an origin for dimensions. These regions are considered perfect for orientation purposes only. A datum is theoretically exact and cannot be used for physical measurements. During machining processes the part is resting against a perfect or exact datum surface. The surfaces of the part that actually rest on these datum planes are the actual datum features, and they contain all the inaccuracies and irregularities. It is from these actual surfaces that measurements are made to check feature relationships. On the drawing, these datum features are identified with the datum feature symbol.

DATUM FEATURE SYMBOL

The datum feature symbol is a rectangular box that contains a letter. Any of the letters of the alphabet may be used except for I, O, and Q, which may be confused with numbers. The letters may be used in any order because alphabetical order is meaningless in this system. Figure 3-4 shows a drawing illustrating the proper symbol and attachment. The important

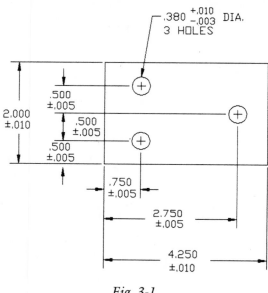

.380 $^{+.010}_{-.003}$ DIA.
3 HOLES

2.000
±.010

.500
±.005

.500
±.005

.500
±.005

.750
±.005

2.750
±.005

4.250
±.010

Fig. 3-1

mental distinction that must be made is that a datum is theoretically perfect while the datum feature itself is imperfect.

The Three-Plane Concept—Flat

Theoretical datum planes or surfaces are established from a perfect three-plane reference frame. This frame is assumed to be perfect with each plane oriented exactly 90° to each other. This reference frame with mutually perpendicular planes provides the origin and orientation for all measurements. These planes are identified as the primary, secondary, and tertiary datum planes.

PRIMARY

The primary datum is the one that is, functionally, usually the most critical feature or surface on the part.

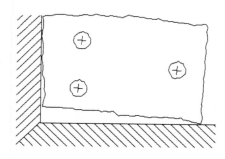

Fig. 3-2

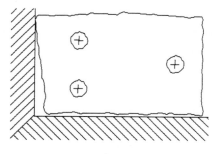

Fig. 3-3

It is typically the largest area when area is involved. The primary datum feature must make contact with the theoretically exact datum plane in a minimum of three points not in a line. The required contact is to prevent the part from "rocking" during manufacturing or inspection processes.

This three-point contact is not difficult to achieve; if the designer has any concern about excessive surface irregularity, a surface control may be specified (see Chapter 6). Figure 3-5 shows an example of the primary datum plane establishment.

SECONDARY

The secondary datum plane must be at a 90° angle to the primary datum plane. The secondary datum feature is usually selected as the second most functionally important feature. This feature must be perpendicular to the primary datum feature. There is only a two-point minimum contact required for this plane. These two points establish the part in the other direction to prevent it from rocking about the primary datum plane. This plane may be a stop, fence, or angle plate on processing or inspection equipment. The

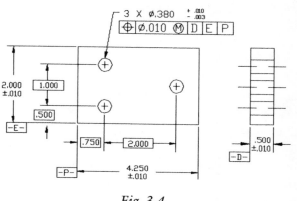

3 X ⌀.380 $^{+.010}_{-.003}$

| ⊕ | ⌀0.010 Ⓜ | D | E | P |

2.000
±.010

1.000

.500

-E-

.750

2.000

-P-

4.250
±.010

.500
±.010

-D-

Fig. 3-4

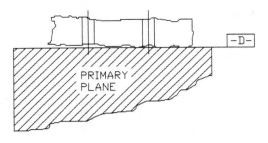

Fig. 3-5

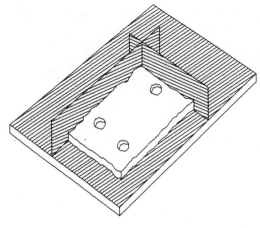

Fig. 3-7

illustration in Fig. 3-6 shows the secondary datum plane.

TERTIARY

The tertiary datum plane must be at exactly a 90° angle to the primary and secondary datum planes. The tertiary or third datum plane is also perpendicular to the other two planes. The part must contact this plane at least at one point. This contact is required for dimension origin and to prevent any back and forth movement in the third plane. The tertiary plane could be a locating or stop pin in a processing or inspection process.

All measurements, set-ups, and inspection are to be made from these three mutually perpendicular planes. Figure 3-7 is an illustration of the theoretical datum reference frame.

The fixture that could be manufactured to check this simple part might look like the one illustrated in Fig. 3-8. The part must contact three points for a primary datum. Often flat parts similar to this may rock if placed on a machine bed or inspection table; the three raised areas will prevent the rocking. The raised areas are not used to check the hole locations. The secondary datum plane requires a two-point contact, which in this illustration are the sides of the pins. The

tertiary datum plane must be in contact with the part in one place only.

Datum Targets

Datum targets are also used to establish the datum reference frame. Datum targets are used to establish orientation on irregular parts, and they may also be used for large surfaces where it would be impractical to use an entire surface as the datum. Usually datum targets are specified on castings, forgings, or weldments or on any other application where it may be difficult to establish a datum plane. Datum targets may be points, lines, or areas of a part that provide an orientation-dimension origin. A datum target symbol and a symbol for each of these three characteristics is used to identify the datum planes.

DATUM TARGET SYMBOL

The datum target symbol is a circle divided in half by a horizontal line. When the symbol is used to

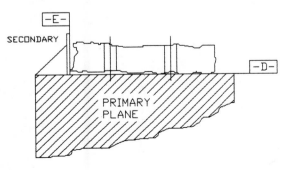

Fig. 3-6

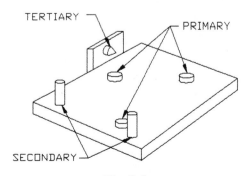

Fig. 3-8

identify a circular target contact area, the top half will contain the diameter (see Fig. 3-9). The contact area diameter is basic. When the symbol is used to identify any other datum target, the upper half remains open. The lower half of the symbol always contains the datum target identification. The identification consists of a letter (the datum reference letter) and target number. Sufficient targets are specified to satisfy the three-plane datum concept and required points of contact (Fig. 10).

Datum target symbols are attached to the datum target with a leader line. A solid leader line indicates the datum target is on the near side of the part; a dashed or hidden line type leader line indicates that the datum target is on the back or far side of the part.

Datum targets are dimensioned with *basic* dimensions or toleranced dimensions. When *basic* dimensions are specified, tooling tolerances apply. Basic dimensions as you learned earlier *do not* have tolerance, but to aid in locating datum targets properly some tolerance is required. Therefore, tooling tolerance (usually one-tenth of specified feature tolerance) is permitted. The feature tolerance used to calculate the tooling tolerance is the specified tolerance for the feature oriented from the datums; see Figs. 3-11 and 3-12. Usually basic dimensions are specified because of subsequent machining and inspection processes.

DATUM TARGET—POINT

A datum target point is identified with an "X" as shown in Fig. 3-12. The "X" or datum target point is usually located dimensionally with *basic* dimensions. The target point is regularly identified in a front view of the part.

Fig. 3-9

Fig. 3-10

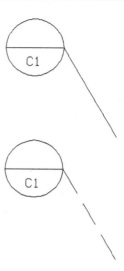

Fig. 3-11

Datum target points are normally simulated or picked up with the point of a cone or spherical radius pin.

DATUM TARGET—LINE

The datum target line may be identified with either a phantom line or an "X." The phantom line is shown in the front view and the "X" in an adjacent view. These symbols are located with basic dimensions as shown in Fig. 3-13.

Datum target lines are simulated with the edge contact of pins.

DATUM TARGET—AREA

The datum target area is identified with a phantom line circle, with section lines inside the circle. If the area is a diameter, then the diameter is specified in the upper half of the datum target symbol. Phantom lines and dimensions are specified to define the size and shape of target areas. Dimensioning of diameters

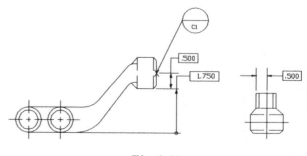

Fig. 3-12

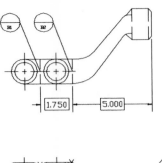

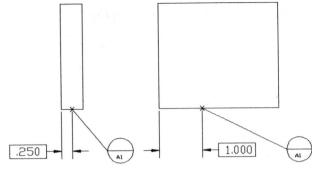

Fig. 3-15

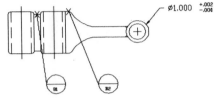

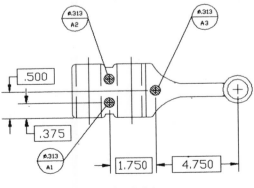

Fig. 3-13

is not required; the datum target symbol specifies the target area diameters. (See Fig. 3-14.)

Datum target areas are simulated with a flat nose pin. In the case of diameters the pin is the diameter specified in the datum target symbol, using gage tolerances for the pin diameter.

DATUM TARGETS—OUT OF DIRECT VIEW

Occasionally it may be necessary to identify a datum target point or line without a direct drawing view. Some design requirements may force the datum identification to other drawing views. The illustration in Fig. 3-15 shows point target dimensioning.

When the datum target line is specified out of the direct view, the locating dimensions are only specified in one direction. In Fig. 3-16 an illustration of the preceding part with a datum target line is shown.

Three-Plane Concept—Circular

Circular parts like noncylindrical parts also require a three-plane concept for repeatable orientation. The primary datum plane is frequently one flat end of the part. Then two planes (X and Y) intersecting at right angles establish the axis. This axis is then used as the theoretically exact datum axis. The two intersecting planes provide dimension origins in the X and Y directions for related part features. Refer to Fig. 3-17.

The drawing in Fig. 3-18 is an example of how the three-plane concept applies to a circular part. Even though the overall diameter of the part is shown as datum "A," only the theoretical axis is used for orientation of related features. Also, basic dimensions are used to locate features from the theoretical planes.

Partial Datums

Occasionally designs require a datum on a particular surface, but not necessarily the entire surface. Examples of such situations would be large parts, weldments, castings and forgings, and plastics. Some designs

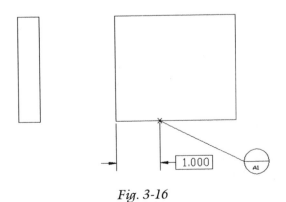

Fig. 3-14

Fig. 3-16

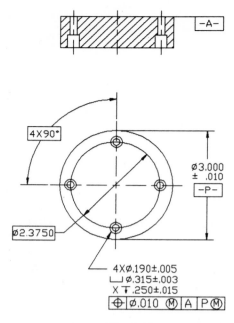

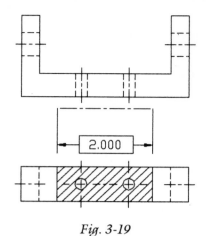

Fig. 3-19

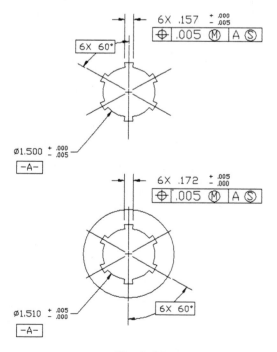

Fig. 3-17

incorporating these parts will have partial datums specified. A partial datum is specified to reduce special treatment to an entire surface such as machining or controlling straightness or flatness. A partial datum is specified with a chain line symbol and cross hatching of the datum area as illustrated in Fig. 3-19.

Fig. 3-18

Datums of Size

A datum of size is any feature subject to size variation based on size dimensions. A feature of size is a hole, slot, tab, pin, etc. A datum feature of size *is not* a single point, line, or plane. Features that are datums and subject to size variation must be verified with a simulated datum (Chapter 5, Rule 5).

In Fig. 3-20 is a drawing of an external feature of size. The diameter of this part is subject to size variation. When features of this type are specified as datums, the material condition must be specified with the datum identification letter in the feature control frame (Chapter 5, Rule 2). The effect of material condition and datum features of size is explained in detail in Chapter 8.

EXTERNAL CYLINDRICAL

External features of size are verified with an adjustable chuck, collet, ring gages, etc. The smallest circumscribed cylinder determines the simulated axis.

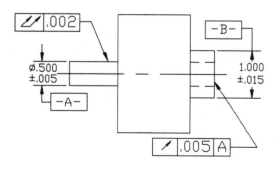

Fig. 3-20

The axis of these irregular features must be established so that measurements and feature relationships may originate from them. The axis of the simulated datum becomes the datum axis for all related dimensions. This axis can be determined with a height gage, coordinate measuring machine, or any other similar instrument. External features are simulated as illustrated in Fig. 3-21.

INTERNAL CYLINDRICAL

Internal features of size are verified in a similar manner. If the feature of size is a hole, the datum axis is determined by the largest inscribed cylinder that will fit the hole. The cylinder must be an expandable pin, mandrel, gage, etc. The simulated axis becomes the datum axis for dimension origins and location of related features. This simulated axis may be determined with a height gage, coordinate measuring machine, or similar instrument.

In Fig. 3-22 is an illustration of how the simulated datum axis is determined.

NONCYLINDRICAL

Internal features may also be specified as datum features. These features are subject to size variation and must be simulated to determine the datum plane, center line, median plane, etc. The simulated datum plane is determined with two parallel planes separated to make contact with the corresponding surfaces of the specified datum feature. The simulated datum may be a gage block, an adjustable gage, or measuring instrument used to establish the datum plane, center line, etc.; measurements then originate from the simulated datum. Figure 3-23 is an illustration of a datum feature simulated with a gage block.

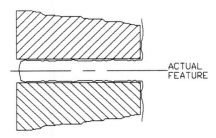

Fig. 3-21

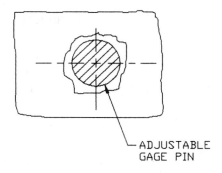

ADJUSTABLE GAGE PIN

Fig. 3-22

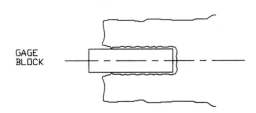

GAGE BLOCK

Fig. 3-23

Datum features of size must also have modifiers specified for them when associated with the positioning of features (Chapter 5, Rule 2). An example of modifiers being specified is illustrated in Fig. 3-24.

When a datum reference letter ("B" in this example) is followed with a modifier, additional consideration must be made for that datum feature. According to the datum/virtual condition rule (Chapter 5, Rule 5), datum feature "B" must be used at its virtual condition even though it is modified to MMC.

Datum reference letters modified with RFS must be treated like any other datum feature of size. The datum is a simulated datum established by an adjustable gage to contact the datum feature as produced.

Summary

Datums are assumed to be theoretically exact in order to ensure repeatability from design to inspection. Datums are dimension origins used to establish measurements and feature to feature relationships. Datum features, on the other hand, are actual part features

Fig. 3-24

that include all variations and irregularities. Datum features may be a point, line, surface, axis, center line, median plane, etc.

Datums are specified to convey the design intent clearly to all that read the drawing. Before datums were specified, assumptions were made about the intent of the design. Today, datums are specified for all parts within a design based on the three-plane concept for both circular and noncircular parts. The three-plane concept provides a solid repeatable orientation.

Datum features are identified with a datum feature symbol or datum target symbol. Letters of the alphabet are used to identify the datum features. Datum features may also be only part of a surface, axis, center plane, etc. If so, the designer will indicate the partial feature with a chain line and required dimensioning for location and length or area of the partial datum. Datums are located either with basic or toleranced dimensions.

Features of size may also be specified as datum features. These features are permitted size variation, therefore requiring adjustable gaging to determine the datum. The gaging provides a simulated datum for dimension origins and feature relationship dimensioning.

Chapter 3 Evaluation

1. Implied datums were open to individual interpretation, therefore the 1982 ANSI Standard requires that datums are _____.

2. Datums are theoretically exact and are used for _____ origins and part _____.

3. List those features of parts that can be used as datums:

   _____   _____

   _____   _____

   _____   _____

4. Is datum reference letter alphabetical order in feature control frames important? _____

5. How many points of contact minimum are required for a primary datum plane? _____

6. The planes of a datum reference frame are assumed to be at _____ degrees basic to each other.

7. Datum target areas are identified with a phantom line _____ with cross hatch lines inside.

8. The upper half of the datum target symbol contains the area _____ when the symbol is attached to a datum target area.

9. Datum targets may be _____, _____, and _____.

10. Datums of size are features associated with a dimension and _____.

11. Datums of size are _____ with adjustable gages, pins, collets, etc.

12. Datums of size also need additional consideration when the _____ are specified with them.

13. Datums are specified on drawings to _____ a clear intent of the design.

14. The minimum point contact required in the three-plane concept is to eliminate part _____.

15. When a datum target area is specified, a _____ nose pin the size of the specified area is required.

Feature Control Frames

Introduction

This chapter deals specifically with feature control frames. A feature control frame is a rectangular box with many compartments. These compartments contain the symbols, tolerances, and datums discussed in Chapters 2 and 3. The symbols, when combined in a specific sequence in the feature control frame, provide a specific control instruction for a feature or group of features to which it is attached. The contents of feature control frames must always be specified in a standard arrangement. Each feature control frame means one setup for manufacturing or inspection. Feature control frames may be single, combined, or composite.

Symbol and Definition

FEATURE CONTROL FRAME

The feature control frame provides a specified control for single or multiple features. The rectangular frame is constructed as required by the designer to control one specific feature or group of features. A feature control frame must contain at least a geometric characteristic symbol and a tolerance value. Feature control frames are read from left to right and line by line in the case of composite control frames. A feature control frame may contain the following symbols and tolerance:

Geometric characteristics

Diameter symbol

Tolerance

Tolerance modifier

Datum reference letter(s)

Datum modifier

Figure 4-1 shows an example of a feature control frame that was specified to control the position of a feature or group of features. The first symbol in the feature control frame is the one used to specify the position or location of a feature(s).

Attachment

INTRODUCTION

Feature control frames may be attached to features in various ways. The method of attachment determined by the designer determines the effect of the control specified for that feature or group of features. Feature control frames may be attached to a surface, axis, or center line. With each method of attachment, the feature control is limited to only that portion of the part or feature to which the feature control frame is attached. For example, if a feature control frame is attached to a surface extension line, then only that surface is controlled.

SURFACE

Feature control frames that control surfaces either control the entire surface or just the surface indicated. For example, round features are controlled all around. The control applies all around because usually it is too difficult to orient the part for control on one side. Generally, a designer will want the same control to apply all around the part. In Fig. 4-2 an example of how a feature control frame is attached to the surface of a circular part is shown.

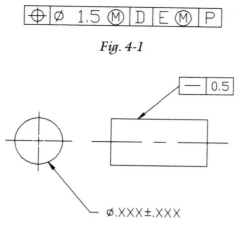

Fig. 4-1

Fig. 4-2

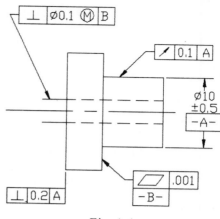

Fig. 4-4

AXIS OR CENTER LINE

Feature control frames associated with round or width-type features are attached to extension lines of that feature. The interpretation means the axis or center line of the feature is controlled with no regard for the feature surface. The designer is specifying the required control so that fasteners will pass through parts or so that parts will mate with each other. The specification of such call outs means the axis or center line orientation is the critical concern rather than the sides of the feature. Examples of how an axis or center line is controlled are shown in Fig. 4-3. (*Note:* Chapter 6 has many examples of feature control frame attachment.)

The feature control frame is usually attached to the controlled feature with one of four methods. These methods are as follows with an example in Fig. 4-4:

The feature control frame is placed below a dimension pertaining to a feature. The leader is from the dimension.

A leader from the feature control frame to the controlled feature.

A side or end of the feature control frame is attached to an extension line from the feature. The feature surface must be a plane.

By attaching a side or end of the feature control frame to a feature-of-size dimension extension line.

Content

GEOMETRIC CHARACTERISTIC SYMBOL

The feature control frame consists of many compartments that contain information specified by the designer. The first compartment of the frame always contains the geometric characteristic symbol for form, orientation, or position.

TOLERANCE

The next compartment always contains a tolerance. The tolerance is either a diameter or a width. If the tolerance is cylindrical, the diameter symbol will precede the specified tolerance. The tolerance is always

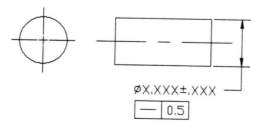

Fig. 4-3

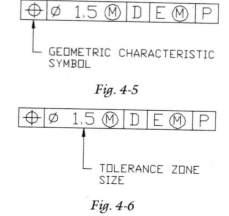

Fig. 4-5

Fig. 4-6

a total tolerance and not a plus/minus tolerance as with the coordinate method of dimensioning and tolerancing.

DATUM REFERENCE LETTERS

When required, the datum reference letter or letters follow the stated tolerance. These letters are not always required, and the number of letters may vary from one to three, depending on the datum reference frame required. The alphabetical order is insignificant; the order from left to right is what establishes the precedence for the datum reference frame. The first letter identifies the primary datum plane, the second letter identifies the secondary datum plane, and the third letter identifies the tertiary datum plane.

MODIFIERS

Modifiers are required in certain situations, or they may be specified in other cases depending on design requirements. Modifier specification is explained in detail in Chapters 5 and 6. The modifier symbols appear in feature control frames as shown in Fig. 4-8.

How to Read Feature Control Frames

LEFT TO RIGHT

Feature control frames are read from left to right. If the feature control frame is a composite or combined symbol, you must read the first line and perform that requirement. Proceed with the next line or lines performing each requirement in succession. Feature control frames do not specify an order of processing, but state the final requirement for the design. In Fig. 4-9 is an example of how to read two feature control frames. Note that a feature control frame is required for each feature or group of features to be controlled. (*Note:* There are conditions that require modifiers to be specified, while in other cases they are

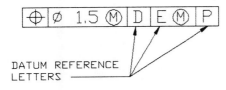

Fig. 4-7

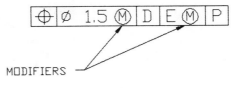

Fig. 4-8

implied. See Chapter 5, Rules 2 and 3, for more information.)

Datum Reference Letter Precedence

THE THREE-PLANE CONCEPT

Datum reference letters are arranged in the feature control frame for a specific orientation of the part. The datum reference letters specify the three-plane concept. The first datum reference letter in the feature control frame determines the primary datum, the second letter identifies the secondary datum, and the third letter identifies the tertiary datum. Feature control does not always require three datum reference letters as shown in Fig. 4-9. The designer will specify the number of letters (datum references) required for proper part orientation and feature control. The datum reference letters are also specified from left to right in their order of precedence; see Fig. 4-10.

TWO DATUM FEATURES

There are some designs that require the identification of two features as datum features to establish a single datum plane, axis, etc. In such cases each feature is identified with a datum reference letter. (See Fig. 4-11.) These two letters then share the same compartment in the feature control frame and are separated by a dash. As illustrated in Fig. 4-11, the center portion of this part is controlled in relation to datum A–B. (The .500 and 1.000 diameter axes establish datum A–B through the part.) These two features establish a single datum axis through the part.

There is one other situation where two letters appear in the same compartment. That situation is when all of the letters of the alphabet are used on one design and there are still more datum features to be identified. In this case a double letter is used to identify a single datum feature. An example of such a feature control frame is illustrated in Fig. 4-12. This ap-

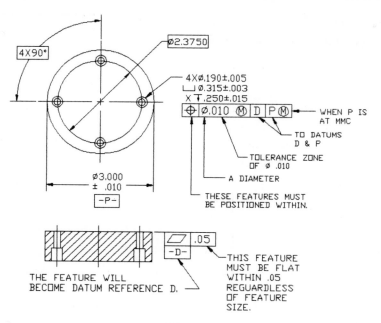

Fig. 4-9

plication is not found too frequently. Usually designs are not complex enough to require such identification.

Types

INTRODUCTION

Feature control frames may be combined in many different ways. Regardless of the combinations used,

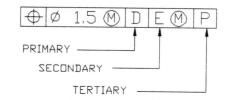

Fig. 4-10

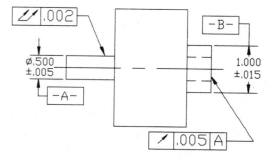

Fig. 4-11

there are only three types of feature control frames. There are those that provide control of a single characteristic; there are many combinations of combined feature control frames; and there are composite feature control frames. Each of these are interpreted one line at a time. Once the first line of a combined or composite feature control frame is used, then the next line comes into effect. The following subsections give some examples of the various combinations that may be created to control the design intent properly.

SINGLE CONTROL

A single feature control frame only provides one control for a feature or pattern of features. The control may be specified to control a surface or to control a feature. Figure 4-13 illustrates two examples of single feature control frames.

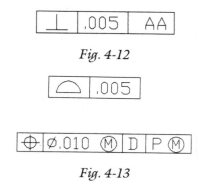

Fig. 4-12

Fig. 4-13

COMBINED FEATURE CONTROL FRAMES

Combined feature control frames are just what their name indicates. There are two or more frames joined together, or there may be a feature control frame and datum reference symbol combined. These symbols, too, are interpreted one line at a time. For example, as shown in Fig. 4-14, a combined symbol that first specified the feature to be parallel to datum "D" and then refines that surface to a flatness of less tolerance.

Another application could be a profile tolerance. Figure 4-15 illustrates a combined feature control frame that specifies a surface control and then refines that control to any line with a lesser tolerance.

Another type of combined feature control frame is where a feature is given a specified tolerance and then identified as a datum feature. This may be common practice where one feature or group of features must be in a specific relationship to another feature. Figure 4-16 illustrates two different combined feature control frames.

COMPOSITE FEATURE CONTROL FRAMES

Composite feature control frames are specified to provide a maximum tolerance to orient or position a feature and then refine that feature to a tighter tolerance. A composite feature control frame will only contain one geometric characteristic symbol; a combined feature control frame may contain two different symbols.

With composite controls, the first line of the feature control frame is considered to be the largest orientation or location tolerance allowable for the feature(s). Then, once the feature is within this tolerance, it is refined to a closer tolerance to control it to ensure assembly with a mating part. An example of a

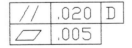

Fig. 4-14

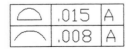

Fig. 4-15

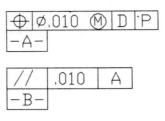

Fig. 4-16

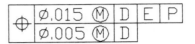

Fig. 4-17

composite feature control frame is shown in Fig. 4-17. This composite control will be explained in depth in chapter eight, and is most frequently specified with location tolerances.

Summary

The feature control frame is specified for each feature or group of features. These feature control frames provide one instruction concerning the form, orientation, or position of features. This means that each feature control frame requires a setup for manufacturing and inspection. The feature control frame contains the information for proper part orientation in relation to the specified datums. The datum reference letters in the feature control frame denote the datum precedence in relation to the three-plane concept.

Feature control frames may be attached to features in one of four ways. The selected method of attachment by the designer controls how features are controlled. The feature control frame may be constructed as single, combined, or composite. Regardless of the type of feature control frame, they all are read from left to right and one line at a time. When one line is read and applied, that line is finished. The information is only used once for a feature.

Chapter 4 Evaluation

1. Feature control frames are rectangular boxes that contain specific information to _____ a feature or group of features.

2. Feature control frames may be single, _____, or composite.

3. Feature control frames may be attached to a _____, axis, or center line.

4. The first symbol in a feature control frame is a _____.

5. The feature control frame must always contain a specified _____.

6. Feature control frames are always read from _____ to _____.

7. When datum reference letters are specified in a feature control, the first letter alway identifies the _____ datum.

8. When reading a composite feature control frame, you always read one _____ of it at a time.

9. Feature control frames consist of a various number of _____ which contain symbols.

10. Is the alphabetical order of the datum reference letters important in the feature control frame? _____

General Rules

Introduction

Geometric Dimensioning and Tolerancing, Y14.5M-1982, like most other standards contains specific rules. This standard identifies three rules, but in fact there are five general rules that apply in various situations. These rules are provided to control some general situations and to provide a common foundation to apply and interpret G.D.T. The rules provide a means to control these situations with only one interpretation throughout the world of industrial drawings. The rules pertain to size and form tolerances, the specification of modifiers, and the origin of datums.

Overview

Most standards have limited "unposted" rules that must be observed at certain times, like driving, for example. G.D.T. also has "unposted" rules that must be observed by designers and those who interpret drawings. We know to drive on the right-hand side of the highway in the United States or that we must drive with lights on at times of impaired vision. G.D.T. also is a standard with "unposted" rules that we must learn and apply when required. These rules, in brief, follow.

Rule One: When only a tolerance of size is specified, that tolerance controls both size and form.

Rule Two: When position or datums of size are specified, modifiers must also be specified.

Rule Three: When form/orientation is specified, the modifier "regardless of feature size" is implied.

Rule Four: For screw threads, splines, and gears, the tolerance and datum reference originate from the pitch cylinder axis.

Rule Five: A virtual condition exists for features of size and datum features of size.

Rule One

Rule one applies to all features controlled with only plus/minus tolerances. Rule one states, "where only a tolerance of size is specified, the limits of size of an individual feature prescribe the extent to which variations in its geometric form, as well as size, are allowed." In such cases the individual feature's size limits control both the amount of variation in size and the form. The actual size of a feature at any cross section shall be within the specified size and plus or minus tolerance. While the feature must meet size requirements at any cross section, the form must also be within these size limits.

The feature surface or surfaces may not exceed the limits of size. These size limits become the boundary of perfect form. For features like pins, the boundary of perfect form would be the nominal size plus the plus size tolerance. This boundary is the features' maximum material condition (MMC) and *no* variation beyond this size is permitted. For internal features, the MMC is the nominal size minus the negative size tolerance. Rule one simply permits variation in a feature's form based on its produced size. Figure 5-1 illustrates a drawing specification.

When the feature varies or departs from MMC toward least material condition (LMC), its form is allowed to vary from perfect. Parts are given tolerance to allow for variation, because the closer to perfect a feature is produced, the more expensive the parts are. Therefore, if variation is allowable, the designer will allow as much as possible based on the design requirements. Features then may wave, bow, taper, step,

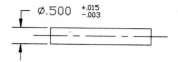

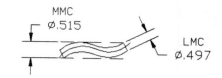

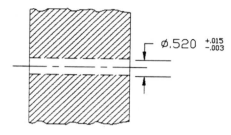

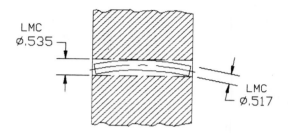

Fig. 5-1

Fig. 5-3

etc., an amount equal to their departure from MMC. Figure 5-2 shows an illustration of feature variation.

When the feature departs from MMC to LMC, there is no requirement for perfect form. The feature is allowed to vary the full limit of the size tolerance. Any variation in the form is acceptable. The variation bow, for example, cannot exceed the MMC boundaries. Figure 5-3 shows an example of how the two mating parts may be produced.

PERFECT FORM NOT REQUIRED

There are certain conditions when rule one is not desired or does not apply. Designers may wish to permit a feature to vary beyond the boundary of perfect form at MMC. In such designs they may add a note, "PERFECT FORM AT MMC NOT REQ'D." This

boundary of perfect form may also be violated by rule five, the datum/virtual condition rule. This rule is discussed in this chapter and in detail in Chapter 7.

Rule one does not control the geometric form of commercial stock, and parts subject to free-state variation. Commercial stock would include bar, sheet, tubing, or structural shapes, or any products produced to established industry or government standards. Parts subject to free-state variation would be rubber and plastic, for example.

FEATURE RELATIONSHIP

Rule one does not control the interrelation of features, it only applies to the individual feature of size. If interrelationship is to be controlled, other geometric controls must be specified. In the example in Fig. 5-4 there is a piece of tubing. The inside diameter is an individual feature completely separate from the outside diameter.

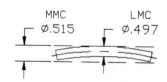

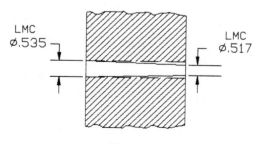

Fig. 5-2

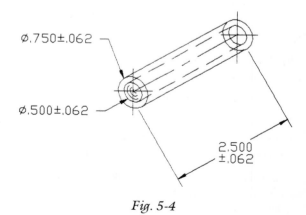

Fig. 5-4

Rule Two

Rule two is primarily a designer's rule. The rule states: "Regardless of Feature Size (RFS), MMC, or LMC must be specified on the drawing with respect to the individual tolerance, datum reference, or both, as applicable." The designer must specify the desired modifier(s) for features subject to size variation when positional tolerances are specified. Features subject to size variation include those being positioned as well as any datum features subject to size variations. An example of such a feature control frame is shown in Fig. 5-5. Various other feature control frames are used in Chapter 8.

Rule Three

Rule three is primarily the interpreter's rule to remember. The rule states: "RFS applies, with respect to the individual tolerance, datum reference, or both, where no modifying symbol is specified." The designer *does not* specify this modifier where applicable. As a drawing interpreter you must remember to apply this modifier. This rule only applies to features subject to size variation. Surfaces of features are not in-

cluded. To read the feature control frame in Fig. 5-6, you must read it RFS even though the modifier is not specified.

Rule Four

Rule four applies to all screw threads, gears, and splines and states that "for each tolerance of orientation or position and datum reference specified for screw threads applies to the axis of the thread derived from the pitch cylinder, for gears and splines the MAJOR DIA., PITCH DIA., or MINOR DIA. must be specified." The designer will specify one of these abbreviations beneath the feature control frame. When there is an exception to the pitch cylinder diameter for screw threads, the designer may specify major dia., pitch dia., or minor dia. In Fig. 5-7 are illustrations of a feature control frame and datum feature symbol with such notation.

Rule Five

Rule five, or the datum/virtual condition rule, applies to datum features subject to size variation. The rule states "depending on whether it is used as a primary, secondary, or tertiary datum, a virtual condition exists for a datum feature of size where its axis or center line is controlled by a geometrical tolerance." Virtual condition is the worst acceptable con-

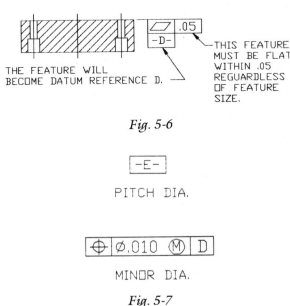

Fig. 5-6

Fig. 5-7

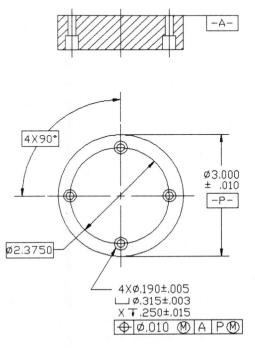

Fig. 5-5

dition of a feature. Datum features of size apply at their virtual condition even though they are referenced in the feature control frame at MMC; see Fig. 5-8. When the designer does not intend for virtual condition to apply for a primary datum, the feature control frame is associated with the size dimension or is attached to an extension of the dimension line.

For secondary or tertiary datum features of size in the same datum reference frame, the size of the sim-

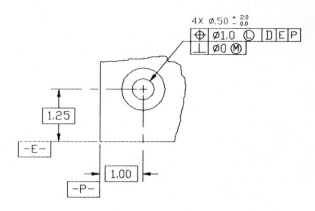

Fig. 5-10

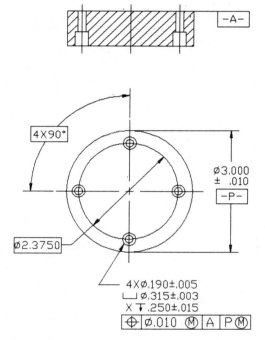

Fig. 5-8

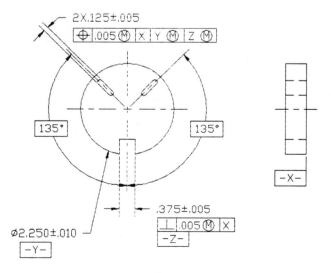

Fig. 5-9

ulated datum is the virtual condition of the datum feature. In Fig. 5-9, datum "Y" is simulated at 2.260 diameter. The axis of this diameter is the simulated datum. Datum "Z" is simulated with a virtual condition width perpendicular to datum plane "X." The center plane of this simulated datum is aligned with datum axis "Y."

Another method of specifying tolerance where the design requirements disallow virtual condition is to specify an appropriate geometric control with a zero tolerance at MMC for features of size. In Fig. 5-10 is an illustration of such a requirement.

Summary

The G.D.T. rules are critical to understand if drawings are to be interpreted properly. These rules provide five guidelines for both the designer and the interpreter of drawings dimensioned with G.D.T. The rules must become part of the interpretation process. It becomes very easy to get involved in the effect of the rules.

Parts may be accepted or rejected incorrectly if some of these rules are ignored. Some rules control form and interrelationship of features, some control the tolerances of form, orientation, and position. One rule controls the datums and virtual conditions of features. These rules must be summarized and remembered in order to interpret drawings properly.

These rules will be applied and referred to in the following chapters. Various examples will aid in remembering their application and effect. The rule number is not always referred to in later chapters when a rule controls a specific application.

Chapter 5 Evaluation

1. Rule one controls both _____ and _____ of features.

2. Rule one only controls _____ features not the interrelationship of features.

3. Rule two specifies that _____ must be specified in the feature control frame when tolerances of position are specified.

4. Rule one specifies a boundary of perfect form at _____ for features of size.

5. When virtual condition is disallowed, _____ positional tolerance may be specified.

6. When geometric controls other than position are specified, which of the modifiers is implied to apply to the tolerance and datum reference if specified? _____

7. Tolerances and datum references originate from the _____ diameter of screw threads.

8. For gears and splines, the designer will specify MINOR, MAJOR, or PITCH DIA. _____ the feature control frame.

9. The basic rules of G.D.T. provide a common _____ to apply and interpret G.D.T.

10. These rules must be memorized because they are _____ specified on the drawing.

Form and Orientation Tolerances

Introduction

Geometric dimensioning and tolerancing provides the designer with a means to define mating parts completely. The portion of the G.D.T. standard that provides methods of controlling part features is the form and orientation controls. These controls are specified for features critical to function and interchangeability where tolerances of size and location do not provide adequate control. When form and orientation tolerances are specified, the tolerance for some features may increase with the use of modifiers.

Remember, G.D.T. is not a replacement for the coordinate system. It is used in conjunction with it to describe completely design requirements. Also remember, that now you have two tolerances to work with. The first, and it should be considered first, is the plus or minus size tolerance. Then geometric tolerances of form, orientation, and location must be considered. You will learn that these tolerances are interrelated in certain applications.

Form tolerances are most frequently applied to single features or portions of a feature. These controls are specified without a datum reference because the features are not controlled in relation to another feature. Orientation tolerances control features in relation to one another. Therefore, a datum reference is required.

Straightness

SYMBOL

Fig. 6-1

DEFINITION

Straightness is the condition where one line element of a surface or axis is in a straight line.

TOLERANCE

Straightness tolerance provides a zone in which a surface element or axis must lie. For surface control the tolerance is implied regardless of feature size (RFS), because the straightness tolerance controls line elements that have no size. For axis control, the tolerance is implied RFS if a modifier is not specified. The tolerance is applied in a view of the drawing where the controlled elements, surface or axis, are shown as a straight line. Each line element or axis must lie within the limits of size for the feature. The tolerance zone may be a width or diameter. The feature must be within the stated size limits at each cross-sectional measurement.

SURFACE CONTROL

When a surface is to be controlled, the feature control frame is attached to the surface with a leader or extension line. In the case of cylindrical features the entire surface must be checked. First, all elements of the surface must be within the specified size tolerance, and then within the limits of the straightness tolerance zone, which is also within the size limits (tolerance). In Fig. 6-2 are examples of how the straightness control may be applied to flat and cylindrical features.

Tolerance Interpretation The tolerance zone is a space between two parallel straight lines that may make contact with the surface of the feature. The tolerance zone for both flat and cylindrical features is applied along the entire surface. This surface may be mea-

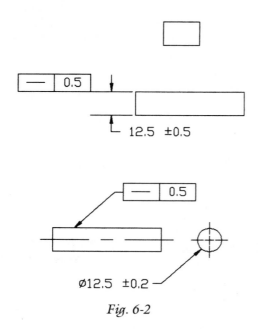

Fig. 6-2

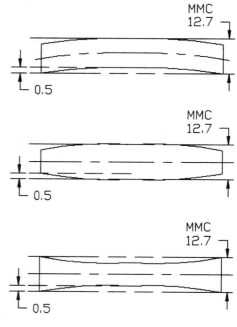

Fig. 6-3

sured or verified with a dial indicator or any other digital readout. The only concern here is how much of the surface has to be verified at one line element at a time. Enough line elements must be verified to ensure that the part or feature is within design requirements. There is no hard and fast rule as to how many measurements to make.

The feature surface may take any shape such as barelling, waving, concaveness, etc., as long as it meets the size requirements, and then falls within the specified form tolerance at full indicator movement (FIM). In Fig. 6-3 are illustrations of how the features may appear. Straightness control specified for a flat part or surface would have a tolerance zone like those shown in Fig. 6-3. The major difference is that for round features the tolerance applies all around, and for a flat feature the tolerance only applies to the surface indicated, as in Fig. 6-2: The tolerance will only apply to the top surface of the part.

AXIS CONTROL

To control an axis, the feature control frame is specified with the diametrical feature size (Fig. 6-4). The control is specified in a view where the axis is shown as a straight line. This type of attachment means that the axis has a specified diametral zone to lie in.

Tolerance Interpretation The axis of the feature may take any form as long as it stays within the diametral

zone. If modifiers are not specified, then the tolerance is implied as RFS. If applicable, maximum material condition (MMC) may be specified for this form control. When MMC is specified, the collective effect of feature size and the straightness tolerance may result in a virtual condition (Fig. 6-5); in other words, the boundary of perfect form may be exceeded to the limit of the stated tolerance.

Since the feature in Fig. 6-4 is a feature of size, the modifier principles do apply if desired by the designer. The feature control frame for the part in Fig. 6-4 could have contained a modified tolerance, Ø0.5 Ⓜ. If the tolerance was specified in this manner, the tolerance zone diameter in which the axis must lie could vary, based on the actual diameter of the part. When the part is produced at 12.7 mm, the tolerance zone is 0.5 mm as specified in the feature control frame. The diametral tolerance zone could increase to 0.9 mm when the part is at the smallest size

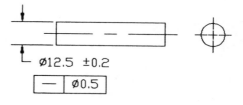

Fig. 6-4

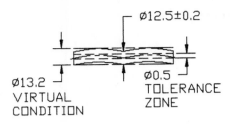

Fig. 6-5

FEATURE SIZE	TOLERANCE ZONE
12.3	0.5
12.4	0.6
12.5	0.7
12.6	0.8
12.7	0.9

Fig. 6-6

or least material condition (LMC). See the tolerance table in Fig. 6-6. This same control may be specified for noncylindrical parts. If straightness is specified for noncylindrical features, the diameter symbol will not precede the stated tolerance in the feature control frame.

Flatness

SYMBOL

Fig. 6-7

DEFINITION

Flatness is the condition of a surface where all elements are in one plane.

TOLERANCE

Flatness tolerance provides a zone of a specified thickness defined by two parallel planes in which the surface must lie. The specified tolerance in the feature control frame is implied as RFS. MMC does not apply to flatness control because only surface area is

controlled and area does not have size. The designer will specify the flatness control in a drawing view where the controlled surface elements are shown as a straight line. Then each surface element must lie within the stated form tolerance zone. The form tolerance zone must be contained within the limits of feature size. Under no circumstances is the feature to exceed the specified limits of size or perfect form at MMC. The feature control frame may be attached to the feature with an extension line of the controlled surface, or attached with a leader pointed to the controlled surface.

APPLICATION

The example in Fig. 6-8 illustrates a proper specification of the flatness form control. The feature is given a size and tolerance that must not be exceeded. Then an additional form control tolerance is applied to the controlled surface. The form tolerance *does not* allow the feature to exceed the specified size requirements.

TOLERANCE INTERPRETATION

Flatness tolerance is the specified distance between two parallel planes of which the upper limit plane must contact the actual feature surface. The other plane then should be the stated tolerance from the first and below all surface area irregularities. The actual surface may be verified with a dial indicator. The indicator should be zeroed for the highest or lowest point on the surface. Then the surface is to be checked suffi-

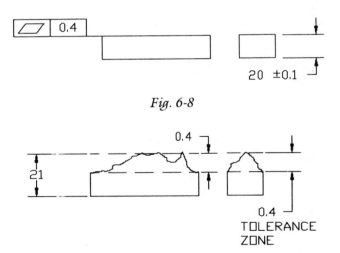

Fig. 6-8

Fig. 6-9

ciently in all directions to ensure that it is within the specified tolerance. The readings obtained are FIM and must not exceed the stated FIM tolerance in the feature control frame.

Circularity

SYMBOL

Fig. 6-10

DEFINITION

Circularity is roundness. Circularity is a condition of a cylindrical surface at any cross-sectional measurement where all points of the surface are perpendicular to and an equal distance from a common axis during one complete revolution of the feature.

TOLERANCE

Circularity tolerance provides a circular zone in which all points of a cross section or slice of the surface must lie. The tolerance zone is two concentric circles the stated tolerance apart. The specified tolerance is implied to be RFS and FIM. Since circularity is a surface control, the modifier principles *do not* apply. The feature control frame is usually specified in the end view. The tolerance zone must be within the size limits of the feature. All surface elements must be within the boundary of perfect form at MMC. This tolerance is not associated with a datum. The surface is controlled or compared to itself, the axis, therefore a datum is not required.

APPLICATION

The circulatory tolerance is applied to compare the circular elements or slices of cylindrical features. Figure 6-11 illustrates the proper application of a circularity tolerance. This tolerance may be specified for any cylindrical feature such as cones, spheres, or cylinders that only require line control around the feature. Circularity may also be specified for internal features that are circular in cross section.

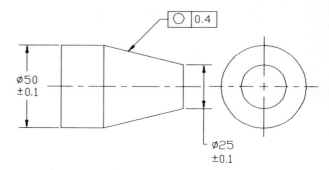

Fig. 6-11

TOLERANCE INTERPRETATION

The circularity tolerance is the space between two concentric circles the stated tolerance apart. Circularity tolerance is a *radial* tolerance. The larger circle must make contact with the actual surface of the controlled external feature. Then the smaller circle would have to be the stated tolerance away from the larger one or the same as the features' smallest permissible diameter. The opposite is true for internal circular features. This tolerance zone is applicable to each cross-sectional element of the feature. The tolerance zone must be perpendicular to the controlled feature axis. All elements of the controlled feature must be within the specified size limits.

Controlled features may be verified with several instruments. The primary concern, however, is how the feature is measured. For example, if a V-block is used, the measurement may include unwanted variables such as lobing, out-of-straightness, the composite effects of a diametral reading, etc. If possible, the measurements should be made in relation to the axis, because the specified tolerance is on the radius. In this way all readings on the indicator will be radial, as the tolerance is intended. Regardless of the method of measurement, sufficient measurements must be made to

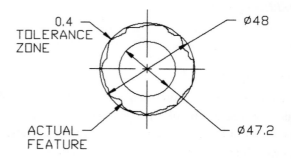

Fig. 6-12

ensure feature acceptance. The specified tolerance is implied FIM for each circular element of the controlled feature. Figure 6-12 illustrates how the tolerance applies to an external feature. (*Note:* The difference between the two diametral measurements is 0.8 or twice the specified tolerance.)

Cylindricity

SYMBOL

Fig. 6-13

DEFINITION

Cylindricity is the condition of an entire feature surface during one revolution in which all surface points are an equal distance from a common axis.

TOLERANCE

Cylindricity tolerance provides a zone bounded by two concentric cylinders in which the controlled surface must lie. Cylindricity tolerance is a *radial* tolerance. The specified tolerance is implied to be RFS in relationship to the feature axis. Since the feature is compared to itself, a datum reference is not required. The tolerance is specified in either drawing view. The feature control frame is attached to the feature with a leader. The cylindricity tolerance zone must be within the limits of feature size. At MMC the feature must have perfect form. Cylindricity tolerance may be applied to internal or external features.

APPLICATION

Cylindricity tolerance is specified in addition to the specified size and tolerance. This cylindricity tolerance remains the same size for all possible feature sizes and is not an addition to the feature size or tolerance. The cylindricity tolerance is a composite control. It controls feature circularity, straightness, and taper. When a cylindricity tolerance is applied, a datum reference is not required because the feature is compared to itself, the axis. In Fig. 6-14 an example illustrates how cylindricity is applied.

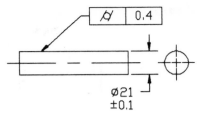

Fig. 6-14

TOLERANCE INTERPRETATION

Cylindricity tolerance is the space between two concentric cylinders the stated tolerance apart. The largest cylinder must make contact with the actual surface of the controlled external feature. Then the smaller cylinder is the stated tolerance away from the large cylinder. Neither of the cylinders are to exceed the feature size limits. When cylindricity tolerance is specified for an internal feature, the cylinders establishing the tolerance zone would be opposite. This tolerance zone controls all points of the surface at one time. The tolerance zone is an equal distance from the controlled feature axis (radial measurement) for the length of the feature.

The controlled feature may be verified with several different measuring devices. This control, however, is composite, and the method of measurement is important. If gaging is not used, then an inspection method must be used that will detect excessive taper, out-of-straightness, and out-of-roundness in relation to the feature axis. The measuring device must not indicate more than the stated tolerance. Figure 6-15 shows an example of an external feature and the specified tolerance zone.

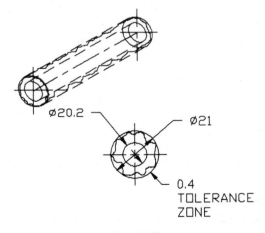

Fig. 6-15

Perpendicularity

SYMBOL

Fig. 6-16

DEFINITION

Perpendicularity is the condition of an entire surface, plane, or axis at a right angle to a datum plane or axis.

TOLERANCE

Perpendicularity tolerance provides a zone defined by two parallel planes, two parallel lines, or a cylinder parallel to a datum. The controlled feature surface, plane, or axis must lie within the specified tolerance zone. Perpendicularity is an orientation control; therefore, a datum reference is required, and the orientation tolerance is implied RFS if not modified to LMC or MMC. Remember, only features of size can be modified, not surface or line features. The feature control frame is specified in drawing a view where the relationship of features is shown.

APPLICATION

Perpendicularity tolerance is specified for designs that require one feature to be perpendicular to another; therefore, a datum reference is required. This datum error is not considered when measuring the controlled feature. Perpendicularity controls all surface error including flatness and angularity. This control is specified in addition to feature size requirements and may be specified in four different methods. A perpendicularity tolerance may be specified for a surface perpendicular to a datum plane, an axis perpendicular to an axis, an axis perpendicular to a datum plane, or line elements of a surface perpendicular to a datum axis. Each of these applications will be explained in detail here.

FEATURE SURFACE PERPENDICULAR TO A DATUM PLANE

This application is the most frequent application for perpendicularity; Fig. 6-17 illustrates this application of perpendicularity. The feature control frame

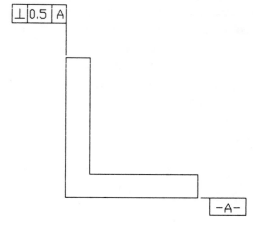

Fig. 6-17

is specified in a drawing view where the relationship between features appears. The feature control frame may be attached to the feature with an extension line or leader.

Tolerance   This perpendicularity tolerance provides a zone defined by two parallel planes the distance of the specified tolerance apart. The tolerance is implied RFS here, because only surfaces are controlled. The tolerance zone must be within the limits of the feature size. All elements of the controlled feature must lie within the perpendicularity tolerance zone.

Tolerance Interpretation   The controlled feature must first meet the size requirements, then the geometric orientation tolerance. The datum reference feature must be placed in contact with the datum plane. Then the tolerance zone must be established at exactly 90° to the datum. All elements of the controlled surface must lie within the parallel planes of the tolerance zone.

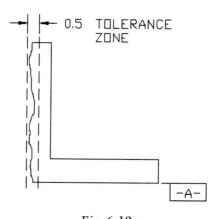

Fig. 6-18

(*Note:* Datum error is not additive to the feature being controlled.)

The actual feature surface may be verified in a number of ways. A simple check can be made with a dial indicator or coordinate measuring machine. The datum feature must be placed in contact with the datum plane. Then with a measuring instrument, make contact with the controlled surface. Zero the measuring device on a high point and continue to measure the entire surface. The FIM is not to exceed the tolerance specified in the feature control frame. Enough of the surface must be measured to ensure that design requirements are met.

FEATURE AXIS PERPENDICULAR TO A DATUM AXIS

Perpendicularity specified in this manner is not a common application for this control. However, this use of perpendicularity does provide the required control for designs where a pin or hole must be at a 90° angle to the axis of another cylindrical part or feature. Figure 6-19 illustrates an application of this orientation control. The feature control frame is specified in a drawing view where the relationship between the datum axis and the controlled feature are shown. The feature control frame is attached to the controlled feature with the feature size call out.

Tolerance The perpendicularity tolerance when specified in this manner provides a tolerance zone for the axis to lie within. The tolerance zone may be a width between two lines or a cylinder. The zone is a width when the diameter symbol is *not* specified in the feature control frame. The specified tolerance is RFS if the designer does not specify a modifier. In the example in Fig. 6-19, the hole is a feature of size; therefore, the modifiers may be specified depending on design requirements. The axis of the feature must be within the boundaries of the tolerance zone if it is a width or diameter at RFS or modified.

Tolerance Interpretation The controlled feature must meet size requirements, then geometric controls can be considered. The geometric tolerance zone must be established at 90° to the datum axis. The controlled feature axis must then be within the width or cylindrical zone as specified in the feature control frame. The feature axis may take any shape—bow, wave, angle—as long as it remains within the tolerance zone. The example in Fig. 6-20 illustrates a width-type tolerance zone. The tolerance applies to the feature as a width in the drawing view shown. If the diameter symbol is specified, the tolerance zone would be cylindrical, allowing the axis of the hole to lean or move in all directions.

The actual controlled hole may be verified with several different methods. This feature may best be verified by using a gage rather than by measuring the angularity. Direct measurements would provide the required information to make this check. If a gage were used for verification, it would simulate the mating part. This gage would have to simulate acceptance of the worst condition for the controlled feature. The worst condition is the collective effect of size and geometric tolerances. This collective effect is known as virtual condition. Rule five controls datum feature "A." Virtual condition is explained and illustrated in Chapter 7.

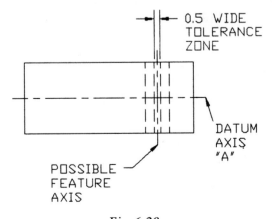

Fig. 6-19   *Fig. 6-20*

FEATURE AXIS PERPENDICULAR TO A DATUM PLANE

Specifying perpendicularity to control the relationship between a cylinder and plane in this manner is a common application of perpendicularity. The feature control frame is attached to the controlled cylindrical feature with the size call out. This call out is in a drawing view where the relationship between features can be seen. Figure 6-21 illustrates a proper specification. The perpendicularity tolerance is usually specified in a drawing view where the relationship is clearly shown.

Tolerance   A perpendicularity specification like this is usually a refinement of location tolerancing. Each control, location, and perpendicularity will have a tolerance specified. The axis of the controlled feature must lie within the tolerance(s). The tolerance zone for this application is usually diametral. Since the controlled features are features of size, the modifiers may be specified depending on the final assembly requirements. The axis of the actual feature must lie within the boundaries of the tolerance zone.

Tolerance Interpretation   The controlled cylindrical feature must meet the size requirements and then geometric controls. If the feature is controlled with a location tolerance and then refined with perpendicularity, the feature must meet the location tolerance and then the perpendicularity tolerance. The perpendicularity tolerance must be at a 90° basic angle to the datum plane. The controlled feature axis must lie within the tolerance zone as specified in the feature control frame or as modified based on actual feature size. The actual axis may be bowed, angled, wavy, etc.,

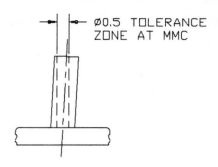

FEATURE SIZE	TOLERANCE ZONE
24.75	1.
25	0.75
25.25	0.5

Fig. 6-22

as long as it remains within the tolerance zone. In Fig. 6-22 is an example of how the actual feature may be produced and how the tolerance may vary when the tolerance is modified to MMC.

This cylindrical feature may be verified with various inspection methods. The most effective way to check the feature with complete assurance that it will assemble with the mating part is with a gage. The gage will have to be made at the virtual condition. A gage at virtual condition will accept all parts that are acceptable for assembly. This gage takes into account all possible feature errors such as out-of-perpendicularity along with out-of-location.

LINE ELEMENT PERPENDICULAR TO A DATUM

The designer may specify a surface perpendicular to another datum plane or axis. The feature control frame is attached to the controlled surface with a leader in a drawing view where the surface to be controlled can be seen. Below the feature control frame the phrase EACH RADIAL ELEMENT must be added by the designer. Figure 6-23 shows an example of how this control is specified.

Tolerance   This perpendicularity tolerance provides a tolerance zone defined by two parallel lines the stated

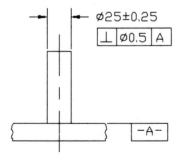

Fig. 6-21

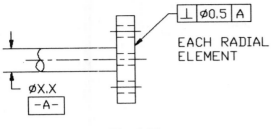

Fig. 6-23

tolerance apart. This zone must be perpendicular to a datum plane or axis. Then each controlled line element must lie between the two lines of the tolerance zone. This tolerance is always RFS because surface elements are being controlled. The tolerance zone must be within the size limits of the controlled feature.

Tolerance Interpretation The controlled feature must first meet the feature size requirements. Then the perpendicularity tolerance is applied by making contact with the actual part surface line by line. The tolerance zone is the thickness of the specified tolerance and perpendicular to a datum. Each line element of the controlled surface is subject to complying with the specified tolerance. Figure 6-24 illustrates how the tolerance zone will appear for a controlled feature.

This feature may be verified with basic precision measurement devices. The controlled feature must be verified at 90° from the datum axis or plane. A simple check with a dial indicator or coordinate measuring machine along line elements of the controlled surface is all that is required. Adequate measurements must be made to ensure design intent. Each line element of the controlled feature must lie within the specified tolerance between two parallel lines. The surface form may vary from measurement to measurement, but each line element must be within tolerance at FIM and RFS.

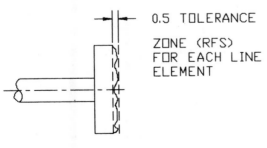

Fig. 6-24

Angularity

SYMBOL

Fig. 6-25

DEFINITION

Angularity is the condition of an axis or plane at an angle other than 90° to another datum plane or axis.

TOLERANCE

Angularity tolerance provides a zone defined by two parallel planes the stated tolerance apart and at the specified basic angle to the datum reference. The controlled feature surface, plane, or axis must lie within this zone. Angularity is another of the orientation tolerances; therefore, a datum reference is always required. The tolerance may be modified if a feature of size is being controlled. If modifiers are not specified, rule three governs the tolerance. The feature control frame is specified in a drawing view where the angular relationship is shown. This relationship must be specified with a basic angle. The datum reference feature irregularities *do not* affect the controlled feature surface or axis.

APPLICATION

Angularity control is specified to control features that are required to be at an angle other than 90° in relation to another feature plane or axis. Angularity tolerance controls surface, plane, or axis errors within the limits of the tolerance zone. Angularity also controls flatness and straightness. The angularity tolerance is in addition to the feature size tolerance. The tolerance provides a zone that controls feature origin in relation to another datum feature. Figure 6-26 illustrates a surface application.

When angularity is specified for an internal feature such as a slot or hole, the tolerance only applies in the view and relative to the datums indicated. The feature is not controlled in any other direction. The tolerance is implied RFS, but may be modified depending on the final requirements of the feature. If angularity is specified for a feature of size, it is usually

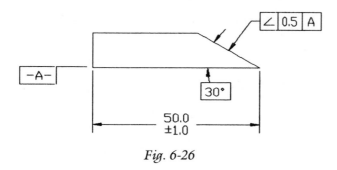

Fig. 6-26

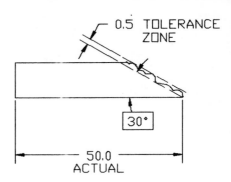

Fig. 6-28

a refinement of a location control. Figure 6-27 illustrates an axis control.

TOLERANCE INTERPRETATION

The controlled feature must meet all other tolerances, and then the orientation control for angularity. The datum reference feature irregularities are not considered when measuring angularity. The tolerance is established at the specified basic angle to the datum. The outer plane of the tolerance zone contacts the highest point(s) on the controlled surface. The inner plane is at the specified tolerance from the first. The controlled surface may take any form through this tolerance zone. The beginning of the angle must be within the length requirement of the part, and may

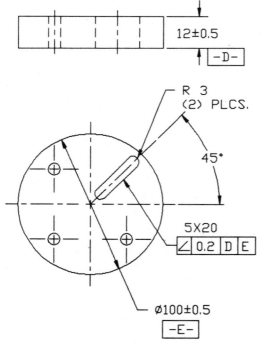

Fig. 6-27

be curved, bowed, or twisted, or at different angle. Figure 6-28 illustrates how a tolerance zone for a surface is applied. The surface may be verified with a dial indicator or coordinate measuring machine.

The control of an axis or plane is slightly different than the control of a surface. The major difference is the tolerance zone. The tolerance only applies in the view that it is specified in and relative to the indicated datums. The actual tolerance zone, then, is a slice the thickness of the specified tolerance that passes through the controlled feature. When viewing the feature as shown in the drawing view with the orientation control, the end of the tolerance would be seen. This zone extends the total length of the part. Figure 6-29 illustrates the tolerance zone for a slot or hole.

Angularity may be verified with several methods. If a modifier were specified, then a gage would be the most effective way to verify correct feature angularity. If the feature is toleranced at RFS, an expandable gage pin, dial indicator, or coordinate measuring machine might be used to determine correct angularity.

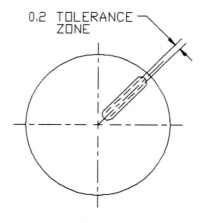

Fig. 6-29

Parallelism

SYMBOL

Fig. 6-30

DEFINITION

Parallelism is the condition of a surface or axis an equal distance at all points from a datum plane or axis.

TOLERANCE

Parallelism tolerance provides a zone defined by two parallel planes, lines, or a cylinder parallel to a datum plane or axis within which the surface elements or axis of the controlled feature must lie. The specified tolerance is implied RFS by rule three. The designer may specify modifiers for features of size based on design requirements for final assembly. The feature control frame is specified in a drawing view where the parallel relationship is shown. This feature control frame must contain a datum reference letter, because parallelism is an orientation tolerance.

APPLICATION

Parallelism tolerance is specified for designs where a plane, surface, or axis must be controlled for parallelism in addition to the feature size tolerance. A datum reference is required with this orientation control. Parallelism also controls flatness of a surface and straightness of an axis within the limits of the tolerance. When parallelism is specified for a surface, the controlled surface must be within the specified size limits. Parallelism orientation may be specified to control three different conditions. The various applications are a surface parallel to another surface, a cylinder parallel to a surface, and a cylinder parallel to a cylinder. These various applications are discussed here in detail.

SURFACE PARALLEL TO ANOTHER SURFACE

This is the most frequent application of the parallelism orientation control. Figure 6-31 illustrates an application. The feature control frame is attached to

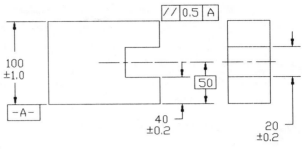

Fig. 6-31

the controlled feature with an extension line in a drawing view where the relationship is shown.

Tolerance   The tolerance for the controlled feature is the space between two parallel planes the specified tolerance apart and parallel to the datum plane. This zone must be within the limits of feature size and is implied RFS according to rule three. All elements of the controlled surface must be contained within the tolerance zone.

Tolerance Interpretation   The controlled feature must first meet the size limit requirements, and then the geometric orientation tolerance. The datum reference feature must contact the simulated datum plane. Any irregularities in the datum reference surface are *not* additive to the controlled feature. The outer tolerance zone plane must be established from the outermost surface elements and must be parallel to the datum plane. The inner plane is then the specified orientation tolerance apart from the outer plane. All controlled surface elements must be within the limits of size and the parallelism tolerance zone. Figure 6-32 illustrates how the tolerance zone would apply to an actual feature.

The surface may be verified with the basic instru-

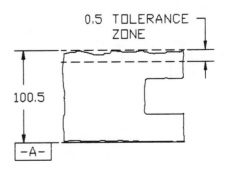

Fig. 6-32

ments of inspection, such as a coordinate measuring machine or dial indicator. The datum surface *must* have the minimum three-point contact with the datum plane. The controlled surface must be measured in many directions and across the entire surface until it is ensured of acceptance.

CYLINDER PARALLEL TO ANOTHER SURFACE

This is a practice that may be specified when a cylindrical feature is to be controlled in only one direction. Figure 6-33 illustrates an application of this orientation control. Here the controlled feature is only controlled in the direction shown in the drawing view that contains the feature control frame. The controlled feature is usually located with a location tolerance and then refined with parallelism.

Tolerance The tolerance for a controlled feature is a width zone the size of the specified tolerance. The axis must lie within two parallel lines the length of the controlled feature. The axis may pass through this zone at an angle or the axis may be curved, waved, etc., as long as it is within the width tolerance zone. This tolerance zone must be exactly parallel to the datum plane.

Tolerance Interpretation To determine if the controlled feature is acceptable, it must first meet the specified size requirements. Then the part must be placed in contact with the specified datum plane. The simulated datum plane may be a knee or stop on the inspection table or work bench. The part *must* be oriented as shown on the drawing. The cylindrical feature is only controlled in the direction it is shown on the drawing. The tolerance zone for the axis to lie within is determined from the simulated datum plane. This can be achieved with direct measurements. The

produced feature must also be measured to determine size and shape. Figure 6-34 shows how the actual feature might appear in relationship to the tolerance zone. (*Note:* The tolerance for this feature may be modified because it is a feature of size. If modified, the tolerance zone would vary depending on actual feature size.)

CYLINDER PARALLEL TO ANOTHER CYLINDER

Specifying a cylinder parallel to another cylinder is one method of controlling cylinders or holes parallel to each other. Depending on the application, this control may provide the required accuracy for final assembly. Figure 6-35 illustrates the application of this parallelism control. The feature control frame is shown in a drawing view where the parallel relationship is required. This parallelism control provides a diametral tolerance zone for the controlled feature.

Tolerance The tolerance zone for the controlled feature may be a cylinder or a width, and modifiers may be specified. The axis of the controlled cylinder must

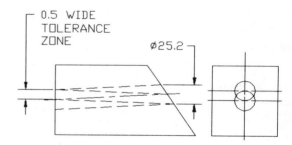

Fig. 6-34

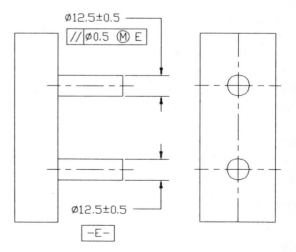

Fig. 6-35

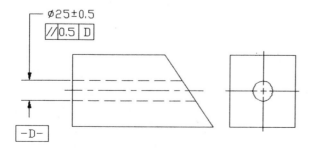

Fig. 6-33

lie within the specified zone for the length of the controlled feature. The controlled feature must be within the limits of size and not exceed the boundary of perfect form at MMC.

Tolerance Interpretation The tolerance may be specified in several ways, as just discussed. Here the tolerance will be discussed as a cylindrical zone at MMC; the datum feature is considered at RFS. Then the datum must be established by the largest circumscribed cylinder. The cylinder should be adjustable so that it will make contact with the irregularities of the pin. The axis of this cylinder is the simulated datum used to verify the controlled feature. The adjustable cylinder may be a gage used for verification.

The controlled feature (pin) has the minimum tolerance zone at MMC that must be parallel to the simulated axis of the datum feature. The axis of the controlled feature must lie within that cylindrical zone. The axis may take any form as it passes through the cylindrical tolerance zone. The tolerance zone for the controlled feature is permitted to increase an amount equal to the amount the feature departs from MMC. See Fig. 6-36 for an illustration of the tolerance zone and the tolerance increase from MMC to LMC.

Verification of this part is best accomplished with a functional gage. The gage would have holes in it that are used to verify the controlled feature. The hole to simulate the datum must be adjustable to *fit* the produced datum feature. The hole to check the controlled feature would be made at virtual condition and be adjustable to *fit* the actual feature.

Another method of specifying the parallelism control for these cylindrical features is to specify both fea-

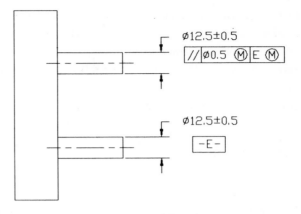

Fig. 6-37

tures of size with the MMC modifier. If they were specified in such a manner, the datum feature is controlled by rule five, the datum/virtual condition rule. By rule five, the datum feature gage hole for simulating the datum would have to be at virtual condition rather than adjustable as it was for RFS. The gage hole for the controlled feature would also be at virtual condition and be adjustable as it was for the other specification. Figure 6-37 illustrates the part described.

Profile

SYMBOLS

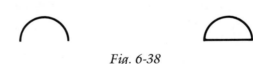

Fig. 6-38

DEFINITION

Profile tolerancing is a method of specifying control of deviation from the desired profile along the surface of a feature.

TOLERANCE

Profile tolerances may be specified either as a surface or line profile. The tolerance provides a uniform zone along a desired true profile (a bilateral zone) or a zone defined by a phantom line either inside or outside the basic true profile of the part (a unilateral zone). The surface of the controlled feature must lie within this zone. The specified tolerance is always RFS, because only a surface is controlled.

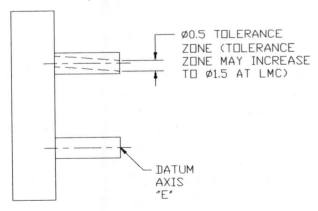

Fig. 6-36

The feature control frame is usually attached with a leader to the surface to be controlled. The feature control frame is specified in a drawing view where the profile is shown. The feature control frame must contain a datum reference and may also have a note added beneath stating "between two points (X and Y)" or "all over."

APPLICATION

Profile tolerance is specified for designs where the surface is to be controlled within a given basic shape. Most frequently, profile tolerances are specified for irregular features that are difficult to control with other form or orientation tolerances. But they may also be specified to control the all around shape of stampings, burned parts, etc. The basic profile of a part is described with basic dimensions, radii, arcs, angles, etc., from a datum(s). Then the specified profile tolerance controls the amount of deviation in relation to the datum reference(s).

Profile tolerances may be specified to control either a surface or line element of a feature. These two applications are discussed next.

SURFACE PROFILE

Surface profile is a method of specifying a three-dimensional control along the entire surface to be controlled. This control is usually applied to parts having constant cross section, surfaces of revolution, weldments, forgings, etc., where an "all over" requirement may be desired. Surface control may be specified to apply only to a limited portion of the surface such as "between X and Y," or it may apply all around. Surface profile may also be specified for coplanar surfaces. Surface profile controls feature size as well as shape. Therefore, feature size and geometric form are verified simultaneously. Figures 6-39 and 6-40 illustrate the proper application of surface profile tolerancing.

Tolerance Surface profile tolerance may be specified as an equal or unequal tolerance on either side of the desired basic profile. If the design requires an unequal application of a bilateral tolerance, a phantom line is included along the basic profile, and the amount of tolerance between the basic profile and the phantom line is specified. Then the balance of the tolerance is

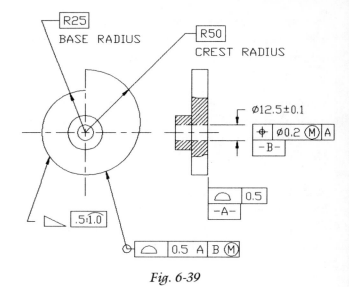

Fig. 6-39

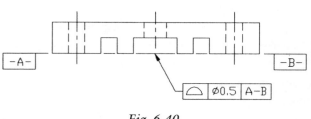

Fig. 6-40

applied to the other side of the basic profile. Phantom lines are also specified when the tolerance is unilateral, all tolerance is applied to one side or the other of the basic profile.

Tolerance Interpretation Profile tolerance for a surface control is implied to be bilateral. If the design requires the tolerance to be unequal or unilateral, phantom lines are added to the drawing view indicating how the tolerance is to be applied. The controlled feature's size and shape are toleranced with the profile tolerance. Surface profile tolerance is a three-dimensional zone. The tolerance applies perpendicular to the true profile at all points along the controlled surface. If the tolerance is bilateral, the actual surface of the feature may vary both inside and outside of the true basic profile. If the tolerance is unilateral, then actual feature may vary only to the inside or the outside of the basic true profile as indicated with a phantom line. Figures 6-41, 6-42, 6-43, and 6-44 illustrate the various tolerance zones permitted by surface profile tolerancing. The tolerance zone is established

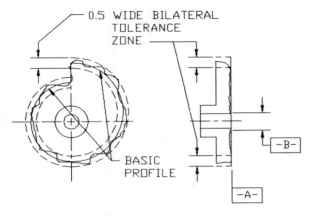

BILATERAL TOLERANCE

Fig. 6-41

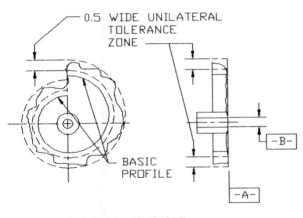

UNILATERAL OUTSIDE

Fig. 6-42

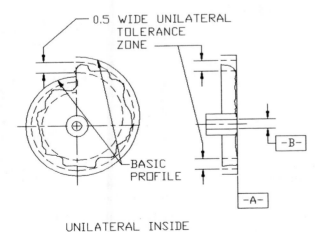

UNILATERAL INSIDE

Fig. 6-43

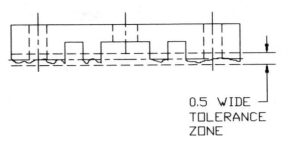

Fig. 6-44

in relationship to the basic true profile of the feature, *not* the actual feature surface.

The actual part surface may be verified with several techniques. The actual part may be compared to a master part, an overlay may be used, optical comparison can be made, or dial indicator or coordinate measuring machine can be used. Verification will depend largely on the accuracy required.

LINE PROFILE

Line profile tolerancing is a method of specifying a two-dimensional control for a single line element along the true profile of a surface. This control is usually specified for the shape of cross sections or cutting planes of parts. The control is most frequently specified for manufactured parts for trucks, automobiles, and marine uses, for items like impellers, body parts, and propellers. This control should be used where "blending" is required. Line profile, like surface profile, may be specified between points or all around. Line profile tolerancing is usually a refinement of some other geometric control, form, and size control. The application of line profile is illustrated in Fig. 6-45.

Tolerance The tolerance for line profile may be specified as either bilateral or unilateral. The bilateral tolerance is implied to be bilateral by the absence of phantom lines to indicate the tolerance zone. The unilateral tolerance zone is indicated with a phantom line either inside or outside the desired true profile of the controlled part.

Tolerance Interpretation The actual surface must lie within the boundaries of the specified zone. The tolerance applies perpendicular to the line profile at all points along the controlled surface. Line profile control may require datum references; when specified, the

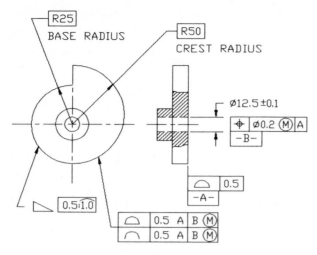

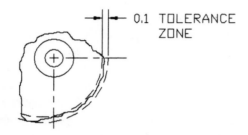

Fig. 6-45

Fig. 6-46

feature must be properly oriented when applying the specified tolerance. The profile tolerance controls both feature size and form. The tolerance zone is established in relationship to the true profile of the controlled feature, *not* the actual surface. All points along the controlled surface must lie within the specified tolerance, as shown in Fig. 6-46.

Line profile may be verified with the same methods used to verify surface profile.

Runout

SYMBOLS

Fig. 6-47

DEFINITION

Runout is a composite form and location control of permissible error in the desired part surface during a complete revolution of the part around a datum axis.

TOLERANCE

Runout tolerances may be specified as either total or circular. The specified tolerance is the deviation permitted in relation to the controlled feature's axis. The specified tolerance provides a zone between two concentric cylinders for total runout control and between two concentric circles for circular runout control. The surface or all points on a cross-sectional line must lie within the specified tolerance zone. The tolerance is *always* implied to be RFS as measured in relation to the datum axis. The tolerance is specified in a drawing view where the controlled feature(s) is shown. The feature control frame is attached to the controlled feature with a leader or associated with the feature size call out. A datum reference is required.

APPLICATION

Runout tolerance is specified for designs where rotation is involved, such as shafts, pulleys, and bearing surfaces. Runout may also be specified for coaxial features; this control, however, is restrictive for manufacturing, because the tolerance is always RFS. Location controls may be a better choice for coaxiality, depending on design requirements. Runout tolerances control the amount of radial deviation for a line or surface of parts that are circular in profile.

TOTAL RUNOUT

Total runout is specified to control feature surfaces that are manufactured with an axis. This control is more stringent than circular runout. It provides a composite control of all surface elements in relation to a datum axis. The controlled surface may be at right angles to or around the datum axis. This control is specified when the composite effect of *all* surface elements together are critical to the final assembly. Figure 6-48 illustrates an application of total runout control.

Tolerance Total runout tolerance is specified in the feature control frame and is implied to be radial, RFS, and FIM. The feature control frame is attached directly to the controlled feature. The actual feature must *not* exceed the boundary of perfect form at MMC.

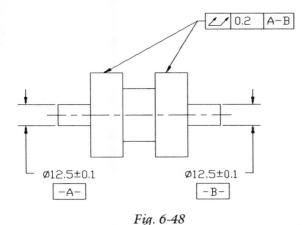

Fig. 6-48

The controlled surface must lie within two concentric cylinders the stated tolerance apart and an equal distance from the datum axis. The entire actual feature must lie within this tolerance zone simultaneously.

Tolerance Interpretation Total runout tolerance *always* applies at RFS. The tolerance establishes a cylindrical tolerance zone the width of the specified tolerance in the feature control frame. This cylinder of tolerance is an equal distance from the datum axis all around. The tolerance is applied simultaneously to all circular and longitudinal elements in one setup during a complete revolution of the controlled feature. The tolerance is composite and cumulative. The feature is controlled for taper, coaxiality, circularity, cylindricity, straightness, angularity, flatness, perpendicularity, and profile. Figure 6-49 illustrates the actual tolerance zone for a controlled feature.

Total runout may be verified by mounting the datum axis in a precision rotational device that will ro-

tate the controlled feature(s) around the datum. The feature may be mounted on a functional diameter or mounted on centers. When the method of measurement is selected, an indicator is set-up in contact with the controlled surface and zeroed. The indicator is *not* rezeroed during the measuring operation. The controlled feature is measured parallel to the datum for circular control and perpendicular to the datum for surfaces perpendicular to the datum. The part must be rotated 360° during the measuring operation, and the indicator reading must not exceed the tolerance stated in the feature control frame. Sufficient measurements must be made to satisfy the drawing requirement.

CIRCULAR RUNOUT

Circular runout is specified to control only surface elements of features that are circular in cross section or surfaces perpendicular to a datum axis. Circular runout is only a line-by-line control of a surface. Each line is completely independent of the other. The tolerance is implied FIM in relation to a feature datum axis. This control is specified when the part function is not critical to rotational speeds. Circular runout is applied in Fig. 6-50.

Tolerance The tolerance for circular runout is specified in the feature control frame and is *always* implied RFS. The tolerance is radial and normally implies FIM. The feature control frame is attached to the controlled feature with a leader or with the feature size callout. The actual feature surface must *not* exceed the boundary of perfect form at MMC. Each cross-sec-

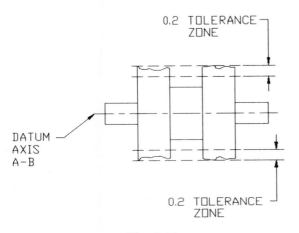

Fig. 6-49

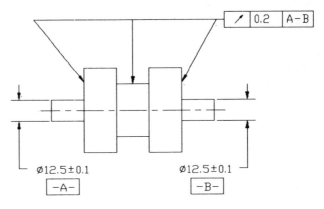

Fig. 6-50

tional slice of the surface must lie within the two concentric circles established by the tolerance zone.

Tolerance Interpretation Circular runout tolerance is a width between two concentric circles the size of the stated tolerance. Each circular element independent of the next must lie within the RFS tolerance zone. The tolerance zone is an equal distance from the datum axis all around. The tolerance controls the cumulative variations in circularity and coaxiality of controlled features around a datum axis. For features perpendicular to a datum axis, the tolerance controls circular elements on the plane. Figure 6-51 illustrates how each circular element is controlled by the specified tolerance.

Circular runout may be verified similar to total runout. The difference is the indicator does not have to be moved along or over the controlled surface. The indicator may be rezeroed for each measurement. The feature must be rotated a complete 360° for each measurement. To ensure acceptability, several independent measurements along the controlled surface should be made.

Unit Control

SYMBOL

There is no symbol for unit control. It is specified as a ratio, usually in the second line of a feature control frame.

DEFINITION

Unit control is a method of controlling feature abruptness in a relatively short length of the controlled feature.

TOLERANCE

The unit control tolerance is usually calculated as a fraction of the form or orientation tolerance. The generally accepted rule of thumb is that the unit tolerance be one-quarter of the stated form or orientation tolerance for any unit of length; that is, it is one-quarter of the overall controlled length of the feature. Unit control tolerancing is usually an additional tolerance to form and orientation control.

APPLICATION

Depending on design requirements, unit control may be specified for features that are controlled with straightness, flatness, and profile tolerancing. The illustration in Fig. 6-52 shows a unit control in conjunction with a profile tolerance. This application specifies a surface profile control of 0.5 mm along the surface. Then units of that surface must also meet the unit control of 0.1 mm per any 56 mm of length.

TOLERANCE INTERPRETATION

The surface profile tolerance is the first consideration in Fig. 6-52. If the controlled feature meets the stated requirements, then unit control is applied. Unit control simply means that the profile tolerance of 0.1 mm applies at any unit of 56 mm of length along the controlled surface. The units of length may be selected at random or even overlap along the controlled surface. Figure 6-53 illustrates how unit control might apply to a surface controlled with a profile tolerance.

Verification of unit control would depend primarily on the application. Unit control specified with a profile tolerance would have to be verified with the

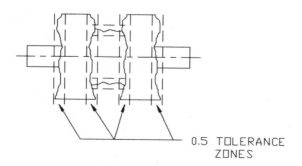

0.5 TOLERANCE ZONES

Fig. 6-51

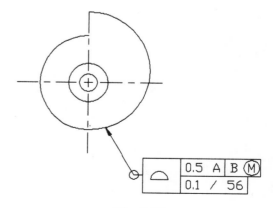

Fig. 6-52

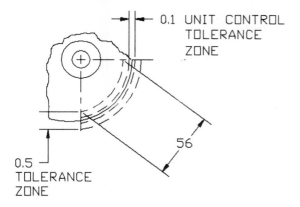

Fig. 6-53

same measuring instruments that are used to verify the profile tolerances. The other applications of unit control can be verified with the instruments used to measure the form or orientation control.

Summary

The G.D.T. form and orientation controls are important to understand, because they control the required shapes of features. In this summary each of the controls will be grouped to aid in remembering what and how they control. The groups will address those controls that control lines and those that control surfaces or areas. All of the controls will be compared to the two basic form controls, straightness and flatness. Straightness only controls one line element at a time or the straightness of an axis. Flatness controls all elements or points of a surface simultaneously in all directions. The remaining controls can all be related to straightness and flatness. Consider circularity and cylindricity. Circularity is a line control around features circular in form. It is like straightness around a feature. Cylindricity controls the complete surface area of circular features. Cylindricity is like flatness rolled around a feature.

Perpendicularity may be compared to both straightness and flatness depending on the features being controlled. Perpendicularity may be specified to control four different types of feature-to-datum relationships. When the control is specified for a plane or surface, the control can be related to flatness. When the perpendicularity control is specified for line element, axis, or median plane, it can be related to straightness.

Angularity is similar to perpendicularity, because it controls all angles other than 90°. Some of the angular relationships will be surfaces, while others will be cylinders or noncylindrical features. Therefore, angularity can be related to flatness when controlling surfaces. When angularity is specified for cylinders or noncylindrical features, a line or median plane is being controlled like straightness.

Parallelism can also be related to either straightness or flatness depending on the type of feature being controlled. When surfaces are controlled, it is similar to flatness where all elements must be in one plane of given thickness. When features with a median plane or axis are controlled with parallelism, the control is similar again to straightness where a line must be straight within a given tolerance.

Profile tolerances are specified to control either surfaces or lines. These may be features that are irregular where flatness or straightness do not apply. Surface profile tolerance provides a thickness zone around or over the controlled surface like a flatness control would provide. Profile of a line is like straightness where only a line element is compared to the true basic profile. Each line element is considered individually.

The runout tolerances also provide for two types of control, where total runout requires measurement of an entire circular surface in one setup similar to flatness. Circular runout is only a cross section or slice of a circular feature control similar to straightness. Only one line element at a time is measured and compared to the specified tolerance.

These comparisons provide a means to identify the control provided by each of the geometric form and orientation controls. There are some differences between controls that should be summarized. The primary difference between a runout control and cylindricity is that runout, either total or circular, controls surface-to-axis, while cylindricity controls axis-to-axis. Another difference between controls is that of perpendicularity and runout. Perpendicularity is primarily specified to control noncylindrical features. Runout is specified to control cylindrical features.

The final thought or explanation that must be made following all of the above discussion is that the design requirements will always dictate the control to be specified. Parts are not designed for themselves, they are designed for function and relationship in a final assembly.

Chapter 6 Evaluation

1. Flatness is a form control that controls surface elements in all _____ within a specified tolerance.

2. Circularity control applies to feature surfaces during one complete revolution as measured _____.

3. Straightness is the condition where one line _____ of a surface or axis is in a straight line.

4. Form and orientation tolerances permit features to vary within the _____ of the tolerance zone.

5. Cylindricity controls the _____ surface of the feature.

6. Per unit control is specified to prevent the continuation of feature _____ or abruptness of the controlled feature.

7. For tolerances of perpendicularity, the zone established by the specified _____ must be within the limits of feature size.

8. The tolerance boundary for a cylindrical feature axis is diametral when the _____ symbol is specified.

9. Feature control frame _____ determines whether a

tolerance is applied to the median plane, centerline, or axis of a controlled feature.

10. Form and orientation tolerances are _____ to be RFS.

11. When a form of orientation tolerance is specified for a feature in relation to a datum feature, the datum feature is _____ to be theoretically exact.

12. With the application of G.D.T., there are two tolerances allowed: _____ and _____.

13. Angularity is the condition of a surface or _____ at an angle other than 90° from a datum.

14. Parallelism is the condition of a surface or axis an equal _____ at all points from a datum plane or axis.

15. For noncylindrical features, angularity tolerance is a _____ and not an angular tolerance zone.

16. Total runout is always implied _____.

17. A _____ tolerance is applied on either side of the basic profile.

18. A _____ profile is specified to control a line element of a surface.

19. Profile tolerance is a method of specifying control of deviation from the desired basic _____ along the surface of a feature.

20. Runout is a composite form and location control of permissible error in the desired part surface during a complete _____ of the part around a datum axis.

Virtual Condition

Introduction

This chapter introduces you to a very important concept concerning the mating parts of an assembly. Today, interchangeability of parts is more critical than ever before. Interchangeability includes those manufacturing situations where subassemblies may be shipped from one country to another for assembly, and also includes the replacement of a single part within an assembly. The required assembly and interchangeability of parts and assemblies can only be ensured when parts are accepted at virtual condition or better. The virtual condition of a part is the condition that defines the boundary of acceptability. This condition is the boundary established by the collective effect of size and geometric tolerance.

Virtual condition is explained and illustrated for both internal and external features. The principles of the modifiers are involved here also. If these are not clear, review them before attempting to comprehend the concept of virtual condition. The virtual-condition concept is especially important when developing gaging. Gages must be made to virtual condition to accept features, but not to accept those worse than virtual condition. Mating parts are dimensioned with consideration for virtual condition. It is through proper dimensioning and gaging with virtual condition that 100% interchangeability and proper function are achieved.

Application

Virtual condition is not a control, but a condition of a feature as a result of size and geometric tolerance. Virtual condition is the boundary at which features are no longer acceptable. This boundary may violate rule one, which in part states "the boundary of perfect form at MMC." In certain applications, the combined effect of the feature size and geometric tolerance will exceed the boundary of perfect form. The following examples will illustrate how the boundary may be exceeded.

EXTERNAL FEATURE

The first indication that a virtual condition may exist for a feature is the size tolerance: it must be a feature of size. The feature must also be controlled with a geometric tolerance. The next indication is whenever a center line or an axis is being controlled. Figure 7-1 illustrates these two conditions. The virtual condition is the MMC size plus the specified geometric tolerance. For this pin, the virtual condition is equal to 12.7 plus 0.5. The virtual condition is 13.2.

The virtual condition of this pin does exceed the boundary of perfect form according to rule one. This is permitted because rule one establishes the boundary of perfect form for only those features that are not controlled with a geometric form tolerance. This pin then may occupy a diametral area up to but not exceeding 13.2 mm. This does not mean that the pin diameter may exceed the stated size and tolerance. If the pin does not meet the specified size requirements, it is not acceptable. It means that with the pin at the largest diameter (12.7) and the axis requiring the entire tolerance zone (0.5), the effective diameter is 13.2 mm. The feature size and geometric tolerance are within specification.

The designer must always consider virtual condition when dimensioning and tolerancing functional parts, because parts are inspected with functional gag-

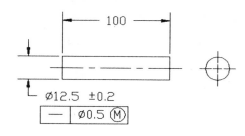

Fig. 7-1

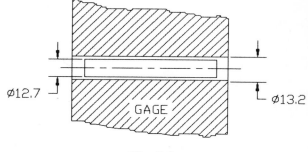

Fig. 7-3

ing at virtual condition that represents the mating part. The functional gage for the pin in Fig. 7-1 is illustrated in Fig. 7-2.

This functional gage diameter must be 13.2 mm in order to check all acceptable pin shapes. This gage is not used to check feature diametral size. Figure 7-3 shows how pin forms are verified with a functional gage. When the pin is at the maximum diameter and at perfect form, it will fit the gage similar to that shown in Fig. 7-3.

When the pin is manufactured at MMC (12.7 mm), the axis is permitted to lie within a diametral tolerance zone of 0.5 mm. This means the pin may also bow or wave to that extent. Figure 7-4 illustrates how this pin would be accepted by the functional gage.

The pin may also be produced to the other size limit (LMC). In this condition the tolerance zone for the axis would increase to 0.9 mm. The pin may take any shape as long as the axis lies within the tolerance zone of 0.9. Figure 7-5 illustrates how pin straightness could vary up to as much as 0.9 mm at the axis and still be acceptable.

PARALLELISM

The concept of virtual condition is not too difficult to follow with one feature. The application on a more

realistic part will help to expand on this concept. The drawing in Fig. 7-6 has two features that must mate with another part, the other part being the functional gage.

The two pins in this part are subject to virtual condition. The lower pin, because it is the datum feature, must be simulated with a gage to establish the datum axis. The upper pin must also be considered with a virtual condition gage, because it meets the requirements of virtual condition.

The gage cylinder for the lower pin would have to be adjustable to fit the produced condition of the pin. The gage cylinder for the upper pin will have a virtual-condition diameter of 13.5 mm. The virtual condition is determined by adding the 0.5 mm tolerance in Fig. 7-1 to the MMC size of 13.2 mm in Fig. 7-5. The gage must be capable of accepting the entire length of the pins. Figure 7-7 illustrates the gage for the part in Fig. 7-6.

INTERNAL FEATURE

Virtual condition for an internal feature must have the same conditions as an external feature. The feature must have both a size and a geometric tolerance. The virtual condition is calculated opposite of an ex-

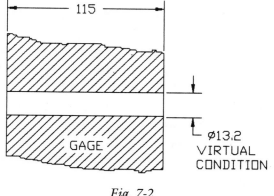

Fig. 7-2

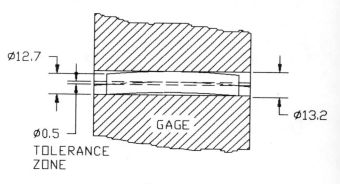

Fig. 7-4

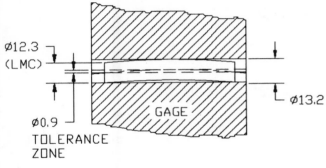

Fig. 7-5

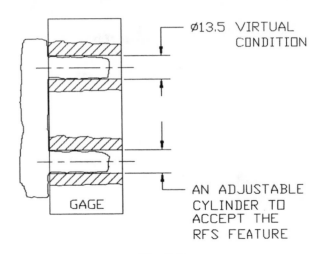

Fig. 7-7

ternal feature. Virtual condition is the feature MMC size minus the geometric tolerance. Figure 7-8 illustrates a feature of size (hole) and the geometric tolerance. The virtual condition for this hole is calculated by subtracting the 0.2 mm geometric tolerance from the 12.4 mm (MMC) feature diameter, to equal 12.2 mm.

The virtual condition of this hole is smaller than the specified size limits of the hole. This is acceptable because the hole at any cross-sectional measurement is within the size limits and the axis may lie any place within the diametral tolerance zone of 0.2 mm at MMC. This hole like the pin in earlier examples is permitted to bow, taper, or wave. The shaft that passes through this hole may not be the same shape as the hole. Therefore, some acceptable condition must be established, the virtual condition. Figure 7-9 illustrates how the hole may be produced.

The functional gage used to check this hole would have a 12.2 mm diameter pin (fixed). The pin is not expandable as it would be if the tolerance was spec-

ified at RFS. The pin is positioned in the fixture so that it would also check the hole in relation to the datums. Figure 7-10 illustrates how this part might fit a gage for checking the hole.

This gage then will only accept parts that are 100% interchangeable if properly designed. The gage accepts holes of any configuration that are within the stated tolerances. This part will need two gages or a gage for the hole and an optical comparison for the profile in order to check and accept the part.

When the controlled feature is also a datum, as is the case with the hole in Fig. 7-8, the gage-pin axis becomes the simulated datum. The gage-pin axis is considered theoretically exact for the purpose of lo-

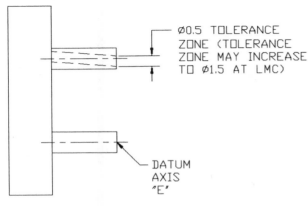

Fig. 7-6

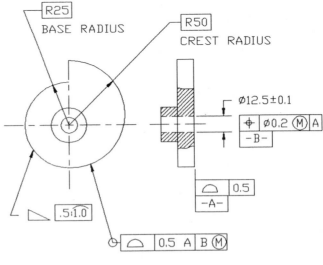

Fig. 7-8

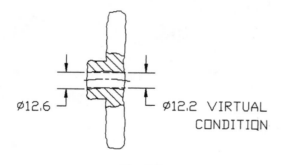

Fig. 7-9

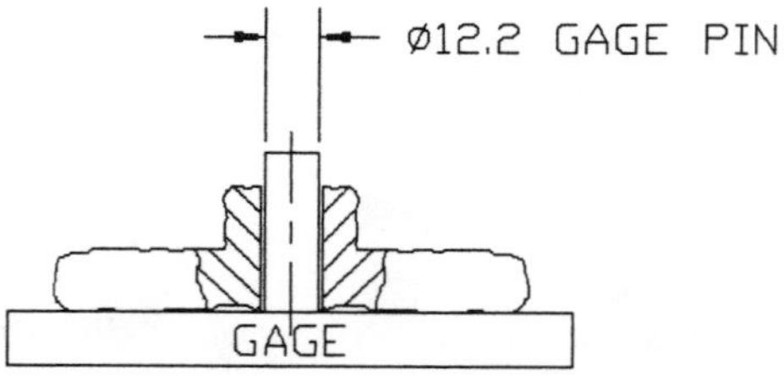

Ø12.2 GAGE PIN

GAGE

Fig. 7-10

cating the part profile. When a gage is used to check a datum feature, the gage axis or center line becomes the simulated axis.

Summary

Virtual condition is a critical concept of G.D.T. from design through inspection. It is with proper application of G.D.T. and interpretation of the virtual condition concept that parts are 100% interchangeable. The concept of virtual condition may best be remembered as an extension of the modifier principles. Virtual condition is always determined to be the MMC size plus the geometric tolerance for external features. In other words the features' largest effective size. For

an internal feature the virtual condition is MMC minus the geometric tolerance. This would be the features' smallest effective size.

Chapter 7 Evaluation

1. Virtual condition is a boundary of perfect form and orientation, generated by the _____ effects of the feature MMC size and applicable geometric tolerances.

2. To determine the virtual condition for an external feature, the feature's _____ tolerance must be added to the feature MMC size.

3. A feature at virtual condition may _____ the boundary of perfect form as required by rule one.

4. Virtual condition is the condition in which _____ are made.

5. Virtual condition is the boundary at which features are no longer _____.

6. Functional gaging at virtual condition represents the _____ part.

7. Virtual condition for an internal feature is calculated the _____ of that for an external feature.

8. Parts and features made at virtual condition or better are 100% _____.

9. According to rule _____, the virtual condition of a feature of size is considered the simulated datum.

10. Virtual condition may also be considered as an extension of the _____ principles.

Tolerances of Location

Introduction

This chapter introduces the principles of tolerances of location. These tolerances or geometric controls are position and concentricity. Symmetrical features are controlled with position. These location controls are specified to control the relationships between features or between features and a datum feature. The relationships are toleranced at the axis, center line, or center plane. Positional tolerance then provides the permissible variation in the specified location of the feature or group of features in relation to another feature or datum. A tolerance of location is applied to at least two features of which one must be a feature of size.

Since one of the features must be a feature of size, the modifier principles do apply. General rule two requires the designer to specify modifiers for all features, tolerances, and datums of size. The advantages of the modifiers can be used to their greatest extent with tolerances of location involving part interchangeability and functionality of mating parts. G.D.T.'s advantages are best realized when position and modifiers are specified.

Concentricity

SYMBOL

Fig. 8-1

DEFINITION

Concentricity is the condition where the axes of all cross-sectional surface elements during one complete revolution are common to a datum feature axis within a specified tolerance.

TOLERANCE

Concentricity tolerance is always implied and specified as RFS, regardless of feature size. The tolerance is a diametral zone in which the axis of the controlled feature must lie. Concentricity is a very restrictive geometric control. A specified tolerance controls the amount of eccentricity error, parallelism of axis, out-of-straightness of axis, out-of-circularity, out-of-cylindricity, and any other possible errors in the feature axis. This tolerance controls all possible errors at the feature axis; therefore, it is difficult to verify and may be excessively expensive. The actual feature axes must lie within the specified tolerance zone.

APPLICATION

Concentricity is considered when critical axis-to-axis control is required for dynamically balanced features. This control is selectively specified, because features may be controlled with runout or position. Runout only controls a surface-to-axis at RFS. Position offers all the advantages of G.D.T. Concentricity is an axis-to-axis control at RFS. Concentricity is normally specified for high-speed rotating parts, rotating mass, axis-to-axis precision, or any other feature critical to function. Figure 8-2 illustrates the proper application of concentricity.

TOLERANCE INTERPRETATION

Concentricity tolerance is always interpreted as RFS. The tolerance zone is diametral around and parallel to the datum axis. Verification is difficult. Concentricity is the determination of the controlled features'

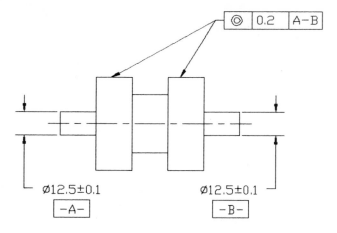

Fig. 8-2

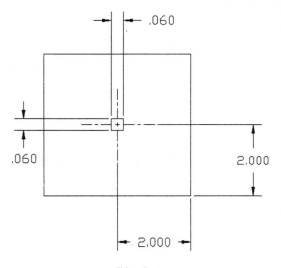

Fig. 8-4

axes in relationship to the datum axis. A radial differential measurement is the most accurate method of determining the actual controlled feature axis. These measurements must be taken opposite of each other. The tolerance zone for Fig. 8-2 is illustrated in Fig. 8-3.

Position

INTRODUCTION

Position is one of the most effective and used controls in G.D.T. This control provides the designer with the ability to specify clearly all design requirements and intentions. Through clearer specification, higher production yields are possible, interchangebility is ensured, and definable quality requirements are attained. The coordinate method does not provide the

same advantages. The coordinate system is not replaced with G.D.T., but is enhanced with it. The two methods of dimensioning and tolerancing may be specified simultaneously on a drawing. As we stated, the coordinate method doesn't provide the same tolerance advantages; this is illustrated in Fig. 8-4–8-8.

Position is always specified for features of size. Datums must be specified. Positional tolerance enables the designer to specify geometric controls that utilize the majority of the advantages for specifying G.D.T. Position may be specified to control feature locations, relationships, coaxiality, concentricity, and symmetry of cylindrical and noncylindrical features of size. Since position is specified for "size" features, the modifier principles must be considered. These principles do

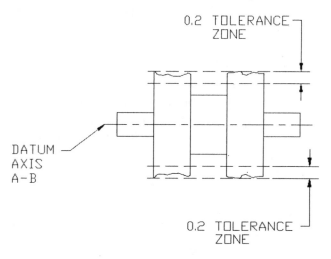

Fig. 8-3

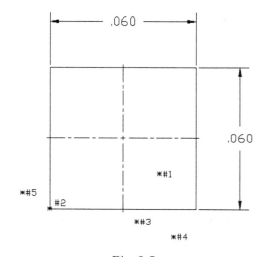

Fig. 8-5

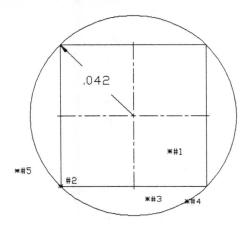

Fig. 8-6

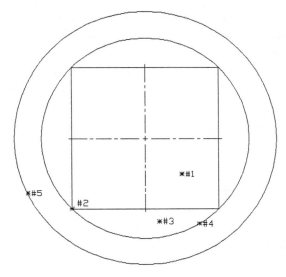

Fig. 8-8

provide many of the advantages for specifying G.D.T. The datum reference frame becomes an important concept when position is specified in relationship to datums. Position is always specified in conjunction with basic dimensions from specified datums or between interrelated features. Basic dimensions establish true position.

Positional tolerance may be explained either in terms of the internal surface of a hole, slot, etc., or in terms of the axis, center plane, or center line. This book, like most documents on the subject, explains position in terms of feature centers. The specified positional tolerance defines a zone within which the center of the feature of size is permitted to vary from true position (theoretically exact) in relation to another feature or datum. True position is established by basic dimensions from specified datum features and be-

tween interrelated features. True position is an axis center line or center plane of a feature as defined by basic dimensions. The tolerance specified in the feature control frame is either a diameter or a width located equally around true position.

The tolerance zone may be modified with the geometric modifiers. Application of the modifiers allows the specified tolerance to increase an amount equal to the actual feature size departure from MMC (maximum material condition) or LMC (least material condition).

Position Theory

INTRODUCTION

Design requirements for assemblies with interfacing mating parts usually relate one feature to another. These features, holes and pins or holes for floating fasteners, relate to each other in 360° of location to each other. The coordinate method of tolerancing *does not* permit 360° flexibility. The coordinate system makes *no* provisions for feature "size" in relation to feature "location." G.D.T. takes into consideration both "size" and "location" when determining feature or part acceptability.

APPLICATION

The following is an explanation of the reasoning behind position versus the coordinate system. For

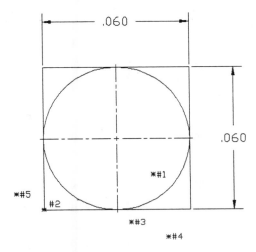

Fig. 8-7

discussion and illustration purposes consider the requirement for a .530 plus .008, minus .002 diameter hole in a part. The hole is to be located within plus or minus .030 to the center of the part. This translates to a .060 square tolerance zone around true position for the actual axis of the hole to lie within. Figure 8-4 illustrates the drawing requirement.

AXIS LOCATION

The axis of the hole must lie within the square zone of .060 to be an acceptable part. To further expand upon this, assume that five parts were inspected, and the axis of each hole is illustrated in relation to the .060 tolerance zone in Fig. 8-5. The illustration shows that two parts are acceptable and three are out-of-tolerance.

DIAGONAL MEASUREMENT

The concept behind positional tolerancing is to use a circular tolerance zone rather than a square one. The circular zone allows for 360° of axis or feature movement. To achieve the circle, the diagonal measurement from center to a corner becomes the radius for a circular zone. In this illustration the diagonal measurement is .042. So, if part number two is acceptable then part three should also be acceptable because it is closer to true position. By specifying a circular tolerance zone, four of the five parts are acceptable. Figure 8-6 illustrates the two tolerance zones.

CIRCULAR TOLERANCE

In Fig. 8-6 the illustrated approach to tolerancing appears to be more logical. But another consideration must be made. The designer intended for the holes to vary in the X and Y directions, not diagonally. Therefore, the circular tolerance zone must not exceed the original intention of .030. The tolerance zone must be an inscribed circle .060 in diameter. Figure 8-7 illustrates the proper comparison between the coordinate and geometric tolerance zones. This illustration appears to have reduced the allowable variation in feature location. It in fact has for a feature of minimum size. When the modifier principles are considered, the tolerance is not lost but gained.

BONUS TOLERANCE

One of the advantages of G.D.T. is greater tolerances based on feature size. In this example, assume parts four and five were produced at the largest permissible size, .538. The preceding .060 tolerance zone has increased to .070. This practice is possible with the application of the modifier principles. The ANSI Y14.5M-1982 standard states in part that "the tolerance is limited to the specified value when the feature is at MMC. Where the actual size departs from MMC, an increase in tolerance is allowed equal to the amount of departure." Figure 8-8 illustrates how the bonus tolerance zone accepts parts four and five when they are made larger than the specified MMC size. All of the parts are acceptable if their axes remains in this .070 tolerance zone for the thickness of the part. This tolerance zone can be considered to be a tube (see Fig. 8-11). The axis of the controlled feature must lie in the tube in any shape or manner as long as it is in there. For noncylindrical features, the zone would be like a box or rectangle.

Position

SYMBOL

Fig. 8-9

DEFINITION

Position is the condition where a feature or group of features is located (positioned) in relation to another feature or datum feature.

TOLERANCE

Location tolerance zones are either cylindrical or noncylindrical; this is determined in the feature control frame. The tolerance zone is cylindrical if the diameter symbol precedes the specified tolerance. The absence of the symbol indicates a width zone. The controlled feature(s) axis or center line must lie within the allowable tolerance zone, which is equally distributed around true position. True position is a theoretically exact location determined with basic dimensions. The specified tolerance zone may increase in size based on actual feature size. Tolerance zone increase is permitted with the specification of modifiers.

APPLICATION

Position may be specified to control nearly all features of a part. Position should be specified whenever the design requirements permit. This control provides an opportunity to utilize many of the advantages of G.D.T. Figure 8-10 illustrates a simplified application of position.

TOLERANCE INTERPRETATION

The tolerance in this example is .030 in diameter when the feature (hole) is at MMC, .528. The tolerance zone is cylindrical because the designer specified the diameter symbol preceding the tolerance. Figure 8-11 illustrates how the tolerance zone is determined. The tolerance zone in this position example

is permitted an increase in size by .010. This is available based on the feature plus/minus size tolerance. The tolerance zone diameter is determined by feature size. Before the feature is produced, only the minimum and maximum tolerance zone sizes can be determined. If this example contained a group of features, each of them may have a different size tolerance zone based on the actual feature size.

Position of Multiple Cylindrical Features

INTRODUCTION

The advantages of G.D.T. and position, in particular, can best be explained when two or more features are positioned. When a group of features are to be controlled with position, a tolerance for location is specified and appropriately modified. Then the feature is related to something, which may be an edge, surface, or other feature(s): These are the datum reference features. Specification in this manner ensures a clearer intent of design. The datum reference frame is utilized for the relationships as required for part function. The rules are applied as required for the specified controls and controlled features. The geometric control—position—brings all of these concepts together better than any of the other controls. An illustration of how multiple features are specified with position is shown in Fig. 8-12.

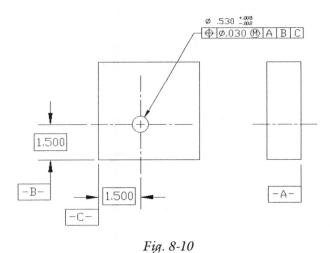

Fig. 8-10

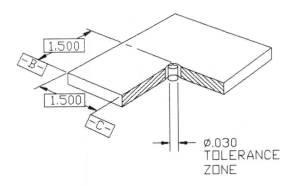

Fig. 8-11

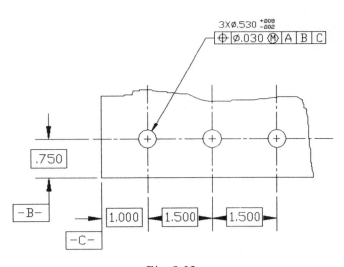

Fig. 8-12

FEATURE LOCATION

The effects of feature size, MMC, and virtual condition can best be illustrated when the explanation begins with the features perfectly located at MMC. Figure 8-13 illustrates these holes with the gage inserted. Note that the gage pins are perfectly centered in the holes when they are in this condition. There is clearance between the pins and holes because the gage is at virtual condition. For this multiple-feature example the gage pins are .498 in diameter. When all features are located perfectly as they are here, there will be an equal distance between each gage pin and the hole. This gage is simulating the fit or assembly of a mating part.

OPPOSITE OFFSET AT MMC

The gage will accept all acceptably located features regardless of their shape, location, and size (must be verified separately). Each feature may be off true position in a different direction and be of different sizes. Figure 8-14 illustrates a possible hole arrangement. These features are at their maximum tolerance offset at MMC. The axis of each feature is .015 of an inch from true position.

The gage will accept this part when the features are offset in various directions. Figure 8-15 illustrates the part and gage. Note there is line contact between the gage and part.

OPPOSITE OFFSET AT LMC

The three features are more likely to be produced at some size larger than MMC. Normally, holes and other features are produced at or near their nominal size or larger. In Fig. 8-16 the outside two holes are

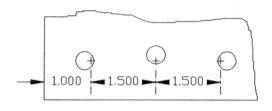

NOTE: HOLES ARE ALL .528 DIA. AND OFFSET.

Fig. 8-14

at their maximum opposite offset at LMC, while the center one is at the maximum vertical offset at MMC. When the holes are produced at LMC as they are in this illustration, the advantage of the modifiers can be realized. That advantage is "maximum tolerances" or bonus tolerance. Here an internal feature is produced at its largest specified size. In the condition the additional feature size increase beyond MMC may be added to the specified location tolerance. Therefore, for the two outside holes, the location tolerance is .040 of an inch, bonus of .010 over MMC.

The gage will accept the part in Fig. 8-16. All features are within specified size and location tolerances. Figure 8-17 illustrates the part and gage. Note that there is a line contact between the gage and features. The examples illustrated in Fig. 8-15 and 8-17 or any other conceivable combination illustrate part acceptance when manufactured to the specifications. The parts are acceptable if they were thin sheet-metal parts or thick castings. The feature axis must lie within the "tube" of tolerance for the thickness of the part.

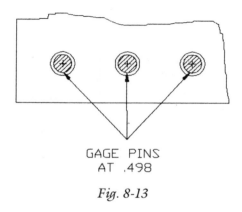

GAGE PINS
AT .498

Fig. 8-13

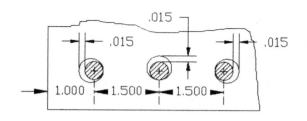

NOTE: HOLES ARE ALL .528 DIA. THE GAGE PINS ARE .498

Fig. 8-15

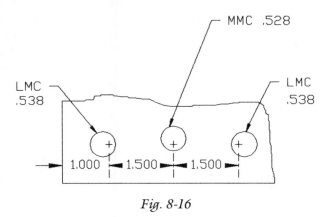

Fig. 8-16

Composite Positional Tolerancing

INTRODUCTION

Composite positional tolerancing is usually specified when feature-to-feature relationship is more critical than the pattern relationship to the part edges. Composite tolerancing is the combining of two or more feature control frames for the purpose of locating a pattern of features to a three-plane datum reference and then controlling the feature-to-feature relationship within the pattern. Composite tolerancing allows the designer the flexibility of specifying the required precision for pattern location and features within the pattern. The modifiers are usually specified to gain the advantages provided by them.

COMPOSITE CONTROL

The combined feature control frame shown in Fig. 8-18 provides composite positional control. The first line controls the location of a pattern in relationship to the specified datum reference frame. This pattern-locating tolerance and datum reference frame provide pattern control for shift or rotation in the part. The second line specifies the tolerance and datum refer-

ence(s) to control feature-to-feature relationship within the pattern. This tolerance allows individual features to vary from true position. The tolerance establishes individual tolerance zones for each feature in the pattern. The tolerance controls feature-to-feature relationship and feature attitude or perpendicularity as it passes through the part. Usually, the same datums and datum precedence are *not* specified for the individual feature control. Generally, only a primary datum reference is specified so that a *separate requirement* is clearly stated.

PATTERN CONTROL

Pattern control is specified to provide an orientation requirement. The control only specifies where the pattern is to be located in relationship to the datum references. The pattern location is specified with basic dimensions from specified datums as illustrated in Fig. 8-19. The features within the pattern are also spaced with basic dimensions. These dimensions provide the *true position* for the pattern and individual features. The specified tolerance of 0.030 Ⓜ applies at the true position for each individual feature in the pattern. This tolerance zone can be thought of as a dart board or target. This is the area that must be hit for the feature to be acceptable. The hit is as good in one place as it is in another, including true position. The designer permits the use of a 0.030 Ⓜ area, use it! If the hit or feature had to be at true position, parts would be excessively expensive. When feature accuracy increases, so does the cost of manufacturing to achieve that accuracy. This type of tolerancing is not to promote ill fitting or poor appearing part and assemblies. If the designer has calculated the tolerance, use what is allowed.

FEATURE CONTROL

The feature control is established by the second line of the combined feature control frame. The tolerance specified here controls the feature-to-datum and feature-to-feature relationship (Fig. 8-20). The specified

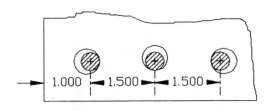

GAGE PINS ARE .498 DIA.

Fig. 8-17

Fig. 8-18

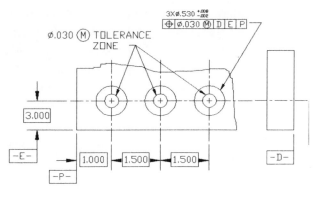

Fig. 8-19

Fig. 8-21

tolerance applies individually to each feature in the pattern and *must* be at the basic dimensions from each other. This feature-to-feature tolerance zone must fall within or at least make a line-to-line contact with the pattern tolerance zone (Fig. 8-21). The axis of each individual feature must lie within this tolerance zone. (*Note*: Example "A" in Fig. 8-21 is not questionable. Example "B" is difficult to verify without a gage and the axis must lie at the point of line-to-line contact.)

AXIS VERIFICATION

The axis of each individual feature must lie within *both* tolerance zones simultaneously for the feature to be acceptable. Figure 8-22 illustrates how the tolerance zones and feature axes may appear. The verification of this pattern is for separate requirements. Each line of the feature control frame would require separate consideration. If the datum references (the letters) and modifiers were the same order of precedence for both lines, then both tolerances would have to be considered in one measurement. By specifying only datum "D," the tolerance zone is perpendicular to that surface, controlling both location and perpendicularity.

Multiple Patterns Located by Basic Dimensions and Related to the Same Datums

INTRODUCTION

Many of the parts throughout the industrial world contain more than one pattern of features. These patterns may be related so that they must be verified together. If the patterns can be separate, they are so specified to reduce the part cost. It is not necessary for the designer to always add the note "SEP REQT." Separate requirements are also specified by different datum references, different modifiers, and different datum order of precedence.

PATTERNS LOCATED RELATIVE TO DATUMS OF NO SIZE

In Fig. 8-23 two patterns of features located from common datums not subject to size variation are illustrated. Since both patterns are located from these datums with basic dimensions, the patterns are verified with one setup or gage. When patterns are specified in this manner, they are *not* allowed any shift or rotation independent of the other pattern.

PATTERNS LOCATED RELATIVE TO DATUMS OF SIZE

When patterns of features are located from the same datums of size, verification again is performed as if

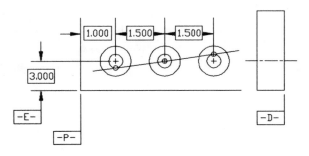

Fig. 8-20

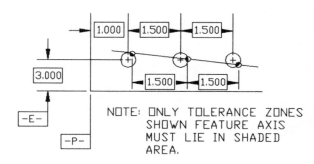

Fig. 8-22

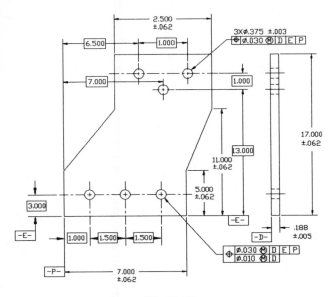

Fig. 8-23

the patterns were a single composite pattern. The feature control frames must contain the same datums in the same order of precedence and with the same modifiers. If the design does not require such an inter-relationship between patterns, the designer may add the notation SEP REQT (separate requirement) under each applicable feature control frame. This notation allows each pattern to be verified independently. It also allows pattern shift/rotation independent of each other. Figure 8-24 illustrates an independent design.

The designer may change the contents of the feature control frames. The designer may designate dif-

ferent datum references, order of precedence, or modifiers thereby relieving the composite pattern requirement. Patterns of features are independent whenever feature control frames are different or when the notation SEP REQT is added beneath the feature control frames.

Patterns Positioned from a Datum of Size

INTRODUCTION

Positional tolerances may be specified to control a pattern of features in relationship to another feature. This type of specification is used when pattern-to-feature relationship is more critical than pattern-to-edge. The single feature in this type of application becomes the locating feature for the pattern. Most frequently, the locating feature is a feature of size. Features of size must have a modifier specified for them in the feature control frame. These locating features are also controlled by rule five, the datum/virtual condition rule. The specification of modifiers in these designs provides the full advantages of G.D.T. If RFS (regardless of feature size) is specified, the tolerances are more restrictive, and verification is more difficult.

REGARDLESS OF FEATURE SIZE

A locating feature (datum) for a related pattern is shown in Fig. 8-25. In this example, the locating datum feature for the pattern is modified to RFS. This

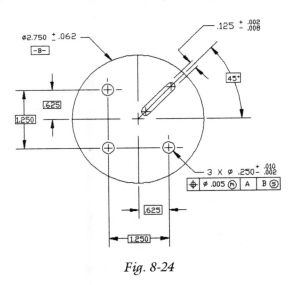

Fig. 8-24

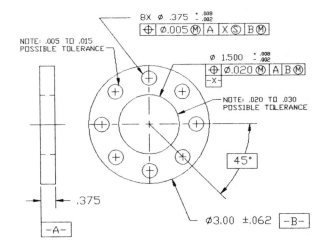

Fig. 8-25

is restrictive, but some designs do require the restriction based on the function of the final assembly.

To begin, the center hole must be located as specified by the basic dimensions and feature control frame. The hole is to be 1.500 in. plus .008 in. or minus .002 in. The hole is allowed a .020 in. diametral tolerance zone for location. In this example the positional tolerance zone for the hole is modified in MMC. This means that the hole has a .020 in. tolerance when it is produced at 1.498 in. If the hole is produced at 1.508 in., the tolerance zone increases to .030 in. The hole was measured to be 1.502 in., so the positional tolerance is .024 in.

Next, the pattern of eight holes must be produced in relationship to the center hole, which has become datum feature "X." Datum feature "X" is modified to RFS in the feature control frame for the eight holes. These eight holes are to be .375 in., plus .008 in. or minus .002 in. They each have a positional diametral tolerance of .005 in. at MMC. Figure 8-26 illustrates the possible tolerance zones for all of the features.

In this example as presented with datum feature "X" at RFS, the pattern shift is restricted to zero. The shift is zero because regardless of the feature location or size, it becomes the dimension origin for the true position of the pattern of eight holes.

Pattern Tolerance When a pattern of features is located from a RFS datum, the pattern is *not* permitted any shift from the datum feature. Here the datum feature axis becomes the origin for the pattern locating

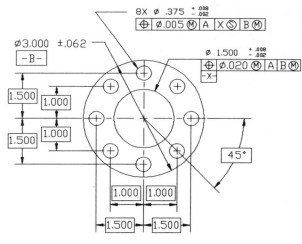

Fig. 8-27

dimensions. Figure 8-27 illustrates the dimensioning for the pattern from datum feature "X." All of these dimensions must be basic.

These basic dimensions establish the true position locations for the pattern. When the basic true position of the pattern in relationship to datum feature "X" is established, the individual tolerance for each feature in the pattern allows variation from true position. Each feature has an individual tolerance based on the actual feature size. This tolerance may vary from .005 to .015 in. because of the MMC modifier following the .005 in. positional tolerance. Each feature in the pattern is permitted to shift or vary as the three features did in Fig. 8-20. Verification of this part would require the establishment of the datum axis. This axis must meet the drawing specifications. After verifying an acceptable datum axis, proper pattern location must be measured. If the pattern is properly located, each feature within that pattern must be verified for proper location and orientation (perpendicularity to datum surface "A"). Verification may be accomplished, depending on required accuracy, with a coordinate measuring machine, paper gaging, or one hard gage that has an adjustable pin to fit the datum feature. The pin must be adjustable because of RFS.

MAXIMUM MATERIAL CONDITION

The MMC modifier specified for a datum feature of size permits the pattern to vary depending on datum size. When the datum feature is at MMC, the pattern variation is restricted. But as the datum fea-

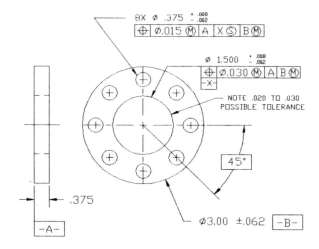

Fig. 8-26

ture increases in size, the amount of pattern orientation shift/rotation also increases. Tolerancing patterns in this manner is more common than RFS. Figure 8-28 specifies positional tolerances at MMC for the same part discussed with RFS.

The center hole or datum feature "X" must be located with basic dimensions. The hole must meet the specified size and location requirements. The hole size may vary from 1.498 to 1.508 in. with a positional tolerance that is .020 in. at MMC to .030 in. at LMC. For this explanation, the hole measures 1.502 in. with a .024 in. positional tolerance.

The pattern of eight holes must be produced in relationship to the axis of datum feature "X." The feature control frame for the eight holes contains datum reference letter "X" with a MMC modifier. This modifier permits some pattern shift/rotation, an amount equal to the feature size departure from MMC. This amount of departure is also the amount the pattern may vary from true position. Figure 8-29 illustrates the available pattern shift/rotation tolerance. The illustration contains a gage pin at virtual condition for datum feature "X." This datum feature is controlled by rule five, datum/virtual condition. The gage pin measures 1.478 in.

Pattern Tolerance The pattern shift/rotation is achieved by the amount of clearance there is between the hole and gage pin. The part may be shifted/rotated all around the gage pin in relationship to the other datum features (part edges).

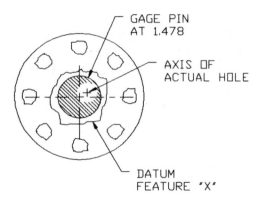

NOTE: PATTERN OF FEATURES ARE LOCATED FROM AXIS OF GAGE.

Fig. 8-29

The eight holes in the pattern establish a true position in relationship to the gage axis. When the pattern orientation is established from the simulated datum axis, each of the eight features has an individual positional diametral tolerance of .005 in. at MMC. This tolerance permits each feature variation based on actual feature size. Figure 8-30 illustrates individual feature tolerance in relationship to the basic pattern orientation.

The MMC modifier permits the pattern some shift/rotation in relationship to the datum feature. Then each individual feature in the pattern is permitted variation from true position. The amount of variation is dependent on the actual feature size. For example, if the datum feature is produced at the virtual condition size, there is no pattern shift/rotation permit-

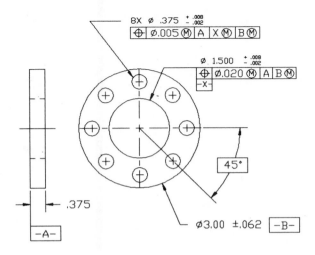

Fig. 8-28

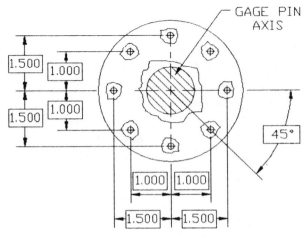

Fig. 8-30

ted. Likewise, if any or all of the eight features were produced at virtual condition, they would have to be located at their true position.

Parts that have patterns of features related to another datum feature(s) may be verified with several methods. Setup time and accuracy required may determine the method of verification. These parts may be verified with a coordinate measuring machine, paper gaging, or hard gaging. Each of these methods provides advantages and disadvantages. The datum feature must be verified first with any method. Separate gaging is required to determine acceptable location in relationship to the datum features. Then from the actual datum feature axis the pattern location is determined. Another gage is required to verify pattern relationship to the datum feature axis.

Zero Tolerancing

INTRODUCTION

Zero positional tolerancing at MMC may be specified where design requirements permit. Additional tolerance then would be allowed when a tolerance is specified in the feature control frame. Zero tolerancing *does not* indicate no tolerance or less tolerance. This method of tolerancing may allow more tolerance and tooling options. The additional tolerance is achieved by the designer adjusting the feature size. The virtual condition for the feature's function is specified as one limit of size with a plus or minus size tolerance depending on an internal or external feature. When the feature is produced at MMC, it must be at true position with zero tolerance.

APPLICATION

To illustrate the advantages of zero tolerancing, the drawing in Fig. 8-12 will be used. In that figure the feature size and positional tolerance is specified as shown in Fig. 8-31.

That specification is a practical specification for a 0.500 in. fastener. The same hole limits of size and

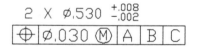

Fig. 8-31

positional tolerance can be achieved with a zero tolerance specification. If zero tolerancing were specified, it would appear as illustrated in Fig. 8-32.

COMPARISON

This method of tolerancing may at first appear as if there is no positional tolerance. The feature at MMC does not have any positional tolerance. However, as the feature departs from MMC, the tolerance zone increases in size from .000 to .040 in. With zero tolerancing, the feature size may vary from .498 to .538 in. In Fig. 8-10 the features could vary in size from .528 to .538 in. with a .040 in. tolerance maximum. Zero tolerancing allows the features to be produced at their minimum functional size or virtual condition and allows for a larger range of tool sizes.

VIRTUAL CONDITION

Zero positional tolerance is based on feature virtual condition. The designer must calculate the virtual condition for parts that assemble, so they may just as well specify that size and provide maximum tolerances for manufacturing. In Fig. 8-33 is a table of

$$2 \text{ X } \varnothing.498 \; {}^{+.040}_{-.000}$$

$$\boxed{\oplus \; \boxed{\varnothing.000 \; \text{\textcircled{M}}} \; \boxed{\text{A}} \; \boxed{\text{B}} \; \boxed{\text{C}}}$$

Fig. 8-32

FEATURE SIZE	POSITIONAL TOLERANCE		DRILL SIZE
.528	.030	= MMC	17/32
.531	.033		
.538	.040	= LMC	
.498	.000	= MMC/V.C	
.500	.002		1/2
.516	.018		33/64
.531	.033		17/32
.538	.040	= LMC	

Fig. 8-33

available drill or punch sizes to produce the features required by the two drawing specifications.

The comparison table in Fig. 8-33 shows the flexibility that is available with zero tolerancing. Also, the lower size limit for an internal feature is both MMC and virtual condition.

Projected Tolerance Zone

SYMBOL

Fig. 8-34

DEFINITION

The projected tolerance zone specifies a required tolerance to prevent interference of mating parts when fixed fasteners are used. The required specified tolerance is projected above the surface that contains the fixed fastener.

TOLERANCE

Projected tolerance zones are usually cylindrical, but may be noncylindrical if the design permits. The tolerance is specified in a combined feature control frame. The upper part of the combined frame specifies the positional tolerance in relationship to the required datums. Then the lower part specifies the height of the tolerance zone above the part that is to contain the fixed fastener.

APPLICATION

A projected tolerance may be specified in an application where one part of an assembly contains the pins or studs to locate another part, cap, or cover to the first. Figure 8-35 illustrates such an application.

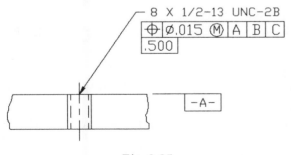

Fig. 8-35

The designer here has specified a projected tolerance of .500 in. above the part containing the fixed fastener. This height is required to ensure assembly of the mating part. The tolerance zone is a .015 in. cylinder 0.500 in. high at MMC.

TOLERANCE INTERPRETATION

The tolerance zone in this example is .015 in. at MMC above the part. This zone is specified to accommodate any error in the drilling or tapping process. The zone permits some misalignment of the two parts or some error in producing the holes. The illustration in Fig. 8-36 shows the two holes produced at slight angles to each other. The axis of the hole in the mating part must be within the specified projected tolerance zone for the thickness of the part (.500 in.). When both parts are produced within their tolerances, the parts will assemble.

Noncylindrical Features

INTRODUCTION

Positional tolerancing may also be specified for noncylindrical and symmetrical features. The feature control frame is an indication that the feature may be noncylindrical: The diameter symbol will not be specified preceding the geometric tolerance. When the diameter symbol is not specified, the tolerance zone is then a width equally divided on either side of the feature center line.

The modifier principles apply to noncylindrical features with the same advantages as cylindrical features. Where modifiers are applicable, the tolerance zones are permitted to increase. The modifiers may also be

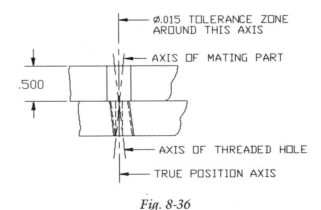

Fig. 8-36

specified for datum features of size. In such applications rule five, the datum/virtual condition rule applies.

APPLICATION

Positional tolerance is applied to noncylindrical features in the same way it is applied to cylindrical features. Positional tolerance and datum references must be specified. The diameter symbol is not specified. The specified feature size and geometric control are usually attached to the controlled feature with extension lines. Figure 8-37 illustrates the application of a noncylindrical position control.

TOLERANCE

The specified positional tolerance of .005 in. at MMC is a total width tolerance for each spline of the feature when the spline is at .313 in. Since these splines are external features, the smaller they are produced, the larger the positional tolerance. In the example illustrated in Fig. 8-38, the .005 in. tolerance is applied equally on either side of the spline center line. The center line may be off to one side, bowed, at an angle, of any other possible form. It must lie within the tolerance zone for the feature to be acceptable. This tolerance controls both the location and the form of each spline.

Bidirectional Tolerancing

INTRODUCTION

Bidirectional tolerancing is another method of positioning round holes. This method of tolerancing is applied when holes may be allowed to vary more in

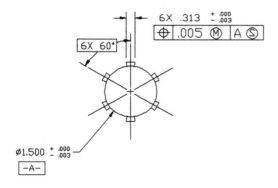

Fig. 8-37

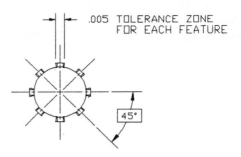

Fig. 8-38

one direction than in another. Holes are located from either rectangular or polar coordinates. The principles apply to either method of coordinate dimensioning. The feature control frame does not contain a diameter symbol before the positional tolerance. The tolerance zone is rectangular and established by two different feature control frames. Bidirectional tolerancing is not limited to holes.

APPLICATION

Bidirectional tolerancing is applied in the same manner as other geometric controls. The feature control frames are attached to the controlled features in a drawing view where the feature relationships can be seen. One feature control frame provides the positional tolerance in one direction while the other one provides control in the opposite direction. Each of the tolerance values represents a distance between two parallel planes or arcs the specified tolerance apart. Figure 8-39 illustrates a rectangular coordinate method or bidirectional tolerancing.

TOLERANCE

The tolerance zones in each direction for bidirectional tolerancing are independent of each other. Combined, they form a rectangular zone. The zones are subject to change when MMC or LMC is specified in the feature control frame. The axes of the holes must lie within the allowable tolerance zones. The axis must remain within the zone for the thickness of the part. If a part is thick enough to where hole angularity or slant through the part is a concern, the designer may specify a perpendicularity or similar control to prevent excessive angularity. The tolerance zone for Fig. 8-39 is illustrated in Fig. 8-40.

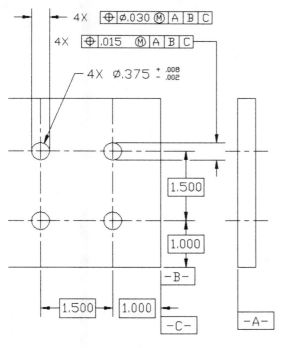

4X ⊕ ⌀.030 Ⓜ A B C

4X ⊕ .015 Ⓜ A B C

4X ⌀.375 + .008 − .002

1.500

1.000

−B−

1.500 1.000

−C−   −A−

Fig. 8-39

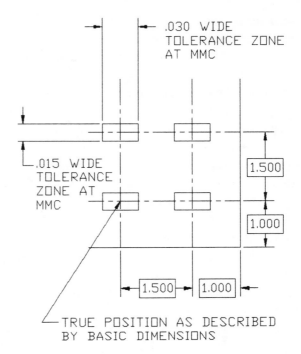

.030 WIDE TOLERANCE ZONE AT MMC

.015 WIDE TOLERANCE ZONE AT MMC

1.500

1.000

1.500 1.000

TRUE POSITION AS DESCRIBED BY BASIC DIMENSIONS

Fig. 8-40

Coaxial Features

INTRODUCTION

Coaxiality control is where the designer is controlling the axis-to-axis relationship of two or more cylindrical feature surfaces simultaneously. Coaxial features share the same principles as any other mating parts. Therefore, datums and modifiers apply. Along with them are the rules governing features of size. Depending on the method of feature control, the advantages of G.D.T. can be realized when controlling coaxial features. The deisgner may specify one of three geometric controls. Depending on design requirements coaxial features may be controlled with position, runout, or concentricity. Their application should be considered in this order.

POSITIONAL CONTROL

Coaxiality is most commonly controlled with position. Position provides the required interchangeability, potentially more tolerance and permits the use of functional gaging. Whenever possible, the MMC modifier should be specified to achieve full advantage of tolerances. Positional control may be specified with the same positional tolerance and datums for both

features, different positional tolerances, and the same datum feature references or different tolerances.

TOLERANCES AND DATUM REFERENCES

When the same positional tolerances and datum references are specified, a single feature control frame is used to control the features. Figure 8-41 illustrates this application. (*Note:* Same tolerances and datums.) When features are controlled in this manner, each of

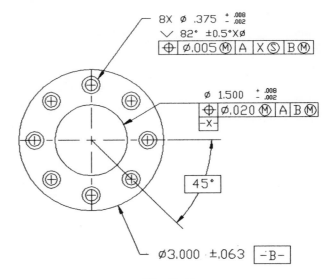

8X ⌀ .375 + .008 − .002

⌵ 82° ±0.5°X⌀

⊕ ⌀.005 Ⓜ A XⓈ BⓂ

⌀ 1.500 + .008 − .002

⊕ ⌀.020 Ⓜ A B Ⓜ

−X−

45°

⌀3.000 ±.063 −B−

Fig. 8-41

the controlled features are produced in accordance with the one feature control frame.

The specified control in Fig. 8-41 requires that both the hole and the countersink axes lie within the same tolerance zone. The illustration in Fig. 8-42 shows how the features may lie within the tolerance zone.

The design may require or allow for different tolerances for the counterbore and hole, but use the same datum reference frame. This application requires two feature control frames: one for the hole pattern and the other for the countersink pattern. Figure 8-43 illustrates this requirement. (*Note:* Different tolerances, same datums.) The two tolerance zones are coaxially located at true position as described by basic dimensions. The feature control frames follow the size specification of the controlled features.

The control specified in Fig. 8-43 requires that the hole pattern lie within one tolerance zone while the countersinks lie within a different coaxial zone. These features as specified also constitute a composite pattern. If the design would allow separate requirements, the designer would have to add the note SEP REQT beneath each feature control frame. Figure 8-44 illustrates the tolerance zones as they were specified.

Designs that require the control of the countersink to the hole will have different tolerances and datums specified to control each group of features. Therefore, each group of features will have a feature control frame specified. In addition, a note is required under the datum feature symbol for the hole and under the feature control frame for the countersink, indicating the number of places each applies individually. Figure 8-45 illustrates this method of application. (*Note:* Different tolerances and datums.)

The method of controlling coaxial features with different tolerances and datums as just described is

similar to specifying control for any two individual features. This specification requires the verification of the hole pattern without regard for the countersinks. When the hole pattern is verified acceptable, then each countersink-to-hole must be verified individually.

This method has two tolerance zones where one is positioned in relation to the axis of each individual hole. In Fig. 8-46 the two tolerance zones are illustrated. The tolerance zone for the countersink is dependent on hole location. Each individual hole axis controls the location of each countersink.

Control of coaxial features with position provides the maximum benefits of G.D.T. All of the advantages of positional control including zero tolerancing may be specified when controlling coaxial features.

RUNOUT CONTROL

Runout control of coaxial features is an axis-to-surface control. Runout is normally specified where a combination of surfaces are cylindrical, conical, or

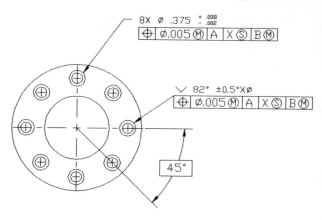

Fig. 8-43

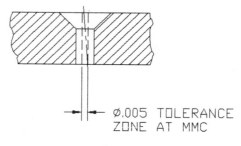

Fig. 8-42

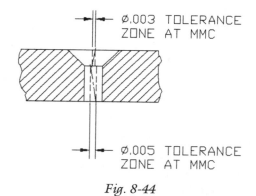

Fig. 8-44

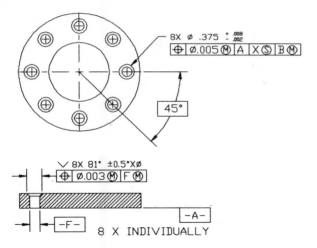

Fig. 8-45

spherical relative to a common datum axis. Runout is more restrictive than position because it is *always* implied as RFS. Runout controls were presented in detail in Chapter 6.

CONCENTRICITY CONTROL

Concentricity controls of coaxiality is an axis-to-axis control of all cross-sectional elements of a surface in relation to a common datum axis. Concentricity tolerance specifies a cylindrical tolerance of a specified diameter at RFS. Concentricity control requires the establishment and verification of all axes irrespective of surface conditions. Concentricity is usually specified as a last resort because it is difficult to verify. Concentricity was presented at length at the beginning of this chapter.

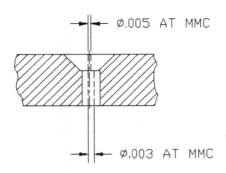

NOTE: AXIS OF C'SINK IS DEPENDENT
UPON AXIS OF HOLE.

Fig. 8-46

Summary

Location tolerancing is the most advantageous part of G.D.T. When location tolerances are properly specified, the rules, modifier principles, datum references, and feature control frames can be applied to their greatest advantages. It is through proper application and interpretation that parts are 100% interchangeable, production yields increase, tolerances are maximum, functional gaging techniques can be implemented, production costs are reduced, among many more advantages.

Location tolerances are the specified amount of variation from true position or another feature. True position is established by basic dimensions from specified datum features. The specified tolerance provides a cylindrical or noncylindrical zone for the depth or length of the controlled feature. This zone is permitted to increase in size as the feature size varies from MMC. The actual feature axis or center line must lie within the zone. The axis or center line may take any form as long as it remains within the zone.

Position is the most frequently applied location tolerance. Concentricity should seldom be specified or it should be limited to features requiring dynamic balance. Position may be specified for nearly all features that require mating part assembly or interchangeability. This chapter should be *studied* to make certain the concepts are understood. Concentration on how the various concepts of G.D.T. are interrelated is important to the reader.

Chapter 8 Evaluation

1. The geometric tolerances of location are _____ and _____.

2. Tolerances of location specify the amount of _____ permitted between features and datum features or between features.

3. Concentricity is always specified or implied _____.

4. The tolerance zone for position is either _____ or _____ as determined from the feature control frame.

5. True position is specified by _____ dimensions.

6. Datum feature references must be _____ on drawings.

7. For an internal feature, positional tolerance at MMC _____ as the feature approaches LMC.

8. When a composite features control frame is specified to locate a pattern of feature in relationship to a datum reference frame, the first line of the feature control frame controls pattern _____.

9. Concentricity tolerance is always specified in relationship to another datum _____.

10. Position tolerance zones extend to the _____ or _____ of the controlled feature.

11. Datum features of size are considered at their _____ size when verifying patterns located from them.

12. Positional tolerances specified for features at _____ lend themselves to functional gaging verification.

13. Position tolerances specified at _____ provide the greatest advantages of G.D.T.

14. Positional tolerances are specified for features of _____.

15. Location tolerancing is a _____ method of controlling part features.

Chapter Nine

Practical Applications

Introduction

This chapter is intended to provide the reader an opportunity to apply various G.D.T. concepts. The examples presented are based on ANSI Y14.5M-1982 practices and should be answered based on that standard. The material consists of a few form and orientation examples, but is primarily centered around positional tolerancing. Position is the most involved of the controls and is probably specified most frequently.

These review problems should be completed after reading the entire text. That way all of the G.D.T. concepts can be applied. This review may be used as a refresher or evaluation of existing G.D.T. knowledge.

Form and Orientation Tolerances

Referring to the drawing in Fig. 9-1, answer questions 1–5.

1. When the feature control frame is specified as $\boxed{-\;|\;.005}$, what is the maximum for D ____ and E ____?

2. When the feature control frame is specified as $\boxed{\square\;|\;.003}$, what is the maximum for D ____ and E ____?

3. Which modifier is implied for these feature control frames? ____

4. Does one of the G.D.T. rules control the modifier application in these feature control frames? ____ If so, which one? ____

5. What is the maximum material condition size of this part? ____

Referring to the drawing in Fig. 9-2 answer questions 6–10.

6. When the feature control frame is specified as $\boxed{\bigcirc\;|\;.005}$, what is the maximum bow, bend, curve, wave, etc., permitted? ____

7. When the feature control frame is specified as $\boxed{\varnothing\;|\;.002}$, what is the maximum bow, bend, curve, wave, etc., permitted? ____

8. If a geometric feature control frame was not specified for this part, is there any control for form? ____ If so, what? ____

9. What is the least material condition size of this part? ____

10. Could the designer specify MMC in these feature control frames? ____ If so, which rule is involved?

Referring to the drawing in Fig. 9-3, answer questions 11–14.

11. The bottom surface of this part is identified as ____ "A".

12. Is the tolerance zone for the .252 hole cylindrical or noncylindrical. Why? ____

13. Is the tolerance zone for the 30° angle a width or an angularly shaped zone along the face of the angle? ____

14. Can any of the geometric controls in this drawing be modified to MMC? ____ If so, which? ____

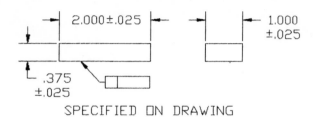

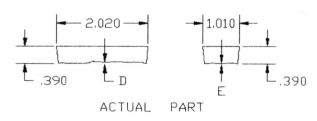

Fig. 9-1

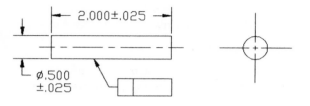

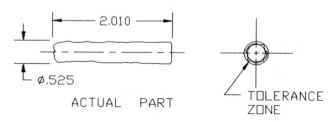

Fig. 9-2

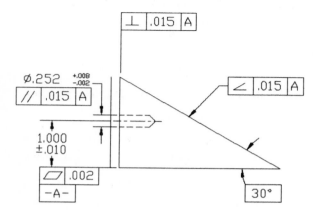

Fig. 9-3

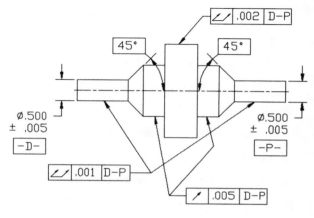

Fig. 9-4

Referring to the drawing in Fig. 9-4, answer questions 15–19.

15. Runout is a _____-to-axis control.

16. Which of the two runout controls specified is more stringent? _____

17. Runout tolerances are always RFS and normally implied to be a _____ reading.

18. All surface elements are controlled simultaneously with _____ runout.

19. The controlled features must meet the requirements of Rule _____.

Referring to the drawing in Fig. 9-5, answer questions 20–25.

20. Concentricity is an _____-to-axis control of two or more features.

21. When concentricity is specified, surface elements are _____ to the functional axis.

22. Concentricity is a _____ control which includes eccentricity, circularity, cylindricity, straightness, etc.

23. The tolerance zone for concentricity is always a _____.

24. Circularity tolerance provides a tolerance zone for the feature _____ to lie in.

25. Circularity is a cross-sectional _____ control.

Referring to the drawing in Fig. 9-6, answer questions 26–30.

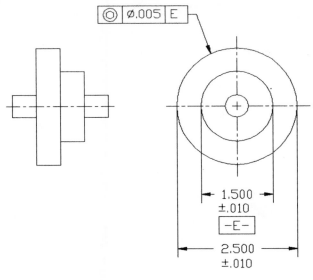

Fig. 9-5

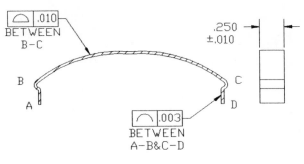

Fig. 9-6

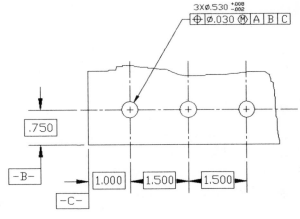

Fig. 9-7

26. Surface profile is a _____ tolerance unless otherwise specified.

27. Bilateral tolerances are assumed to be divided equally on either _____ of the profile unless the designer specified something else.

28. Line profile tolerances are usually specified as a refinement for _____ profile tolerances.

29. What is the total amount of variation permitted with the surface profile tolerance specified? _____

30. Profile tolerances are usually specified for _____ or _____ surfaces.

Hole Pattern—Single Control

Referring to the drawing in Fig. 9-7, answer questions 31–38.

31. What is the hole tolerance at LMC? _____ MMC? _____

32. How close together can two of the holes get? _____

33. Must the axis of the hole always be within the allowable tolerance zone for that hole? _____

34. Datum "A" controls the _____ or _____ of the holes as they pass through the part.

35. What determines the location of this pattern? _____

36. Must each hole tolerance zone be at the specified basic dimension from each other? _____

37. What is the hole size at MMC? _____ LMC?

38. What size would the gage pins be if a gage were made to check these holes? _____

Hole Pattern—Composite Control

Referring to the drawing in Fig. 9-8, answer questions 39–47.

39. The _____ tolerance is 0.30 diameter at MMC.

40. Each hole in the pattern is controlled with a _____ diameter tolerance at MMC.

41. Each hole in the pattern _____ shift or move within the pattern the specified tolerance.

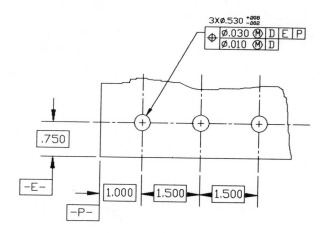

Fig. 9-8

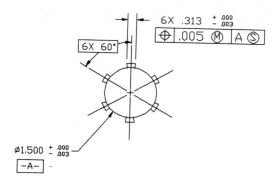

Fig. 9-9

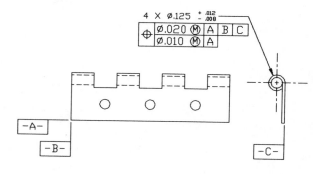

Fig. 9-10

42. The axis of each feature within the pattern _____ be the basic dimension apart.

43. Each feature within the pattern is permitted _____ tolerance at MMC.

44. The .010 zone _____ be within or at least make line to line contact with the .030 zone.

45. The .010 tolerance is specified to _____ feature-to-feature relationship closer and to control feature _____ .

46. The actual feature axis for each hole must simultaneously _____ through both tolerance zones.

47. The feature tolerance zone could range in size from _____ to _____ depending on actual feature size.

Noncylindrical Features

Referring to the drawing in Fig. 9-9, answer questions 48–52.

48. What is the tab MMC? _____

49. What is the shape of the tolerance zone? _____

50. Does the tolerance control the amount tab angularity in relation to the part axis? _____

51. Why is datum "A" modified in the feature control frame? _____

52. Does the basic dimension, 6 places by 60 degrees, have a tolerance? If so what is it? _____

Coaxial Features

Referring to the drawing in Fig. 9-10, answer questions 53–57.

53. This feature control frame is referred to as a _____ feature control frame.

54. The .020 tolerance provides a _____ tolerance zone.

55. The pattern coaxiality is held to _____ at MMC.

56. The .010 tolerance controls _____ feature in the pattern.

57. Why are the modifiers left off the datum reference letters? _____

Past Practices

This Appendix is provided to illustrate two major past practices. There are many changes and updates to the standard as we know it today; most of these changes are simplifications of various symbols. Those past practices are fairly straightforward and understandable if found on today's drawings. There are two practices, however, that may not be as well understood. These practices are symmetry and plus/minus positional tolerancing with implied datums.

Symmetry

SYMBOL

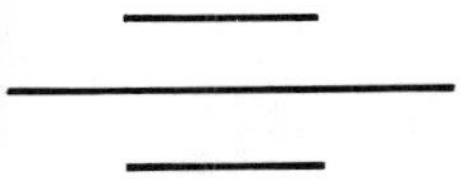

Fig. A-1

DEFINITION

Symmetry is the condition where a feature or part has the same profile on either side of the center line (median plane) of a datum feature.

TOLERANCE

The tolerance for symmetry is always implied to be RFS. The tolerance is applied equally on either side of the controlled feature center line. The implied modifier restricted the tolerance to the specified amount only.

APPLICATION

Symmetry was specified for features to be located symmetrically with respect to the median plane of a datum feature. It may have also been specified for a feature in a common plane with a datum plane. This symbol was specified extensively throughout industry. Figure A-2 illustrates an application of symmetry.

TOLERANCE INTERPRETATION

The specified tolerance established a tolerance zone the specified tolerance in width, with half of the tolerance on either side of the datum feature center line. This width zone allowed the feature to vary from side-to-side or angularly within the tolerance zone. The tolerance zone size was not permitted to vary with feature size. Figure A-3 illustrates how the interior of the part could vary in relation to the exterior, which is the datum feature. The controlled feature is permitted a maximum of .005 in. shift to the side in either direction.

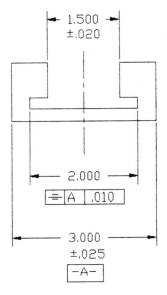

Fig. A-2

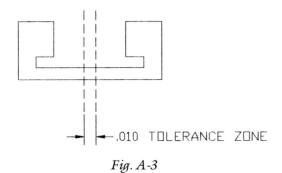

.010 TOLERANCE ZONE

Fig. A-3

Plus/Minus Positional Tolerancing

Many existing drawings throughout industry have features and patterns of features located with plus/minus tolerances following dimensions from implied datums. Then features within the pattern were located with basic dimensions and positional tolerances. This combination method will be explained and illustrated below. The part used is the same as that in Fig. 8-23, where composite positional tolerance was illustrated and explained.

SYMBOL AND DEFINITION

The position symbol and definition are the same for the current practice as they were for the previous practice. Refer to Fig. 8-9 for the symbol and the paragraph following the figure for the definition.

APPLICATION

The drawing that will be used to explain and illustrate the combination of positional and plus/minus tolerancing is the same as Fig. 8-23 with the dimensioning changed to the 1973 standard, as illustrated in Fig. A.4. Note that the plus/minus pattern-locating dimensions are not from specified datums. Dimension origins were implied by the surfaces the dimensions originated.

PATTERN-LOCATING TOLERANCE

The plus/minus tolerance created a square pattern-location tolerance zone twice the size of the plus/minus tolerance. The plus/minus tolerance in this example is .010 in. The square tolerance zone would measure .020 in. in the X and Y directions. Refer to Fig. A-5 for an illustration of the pattern positional tolerance zone.

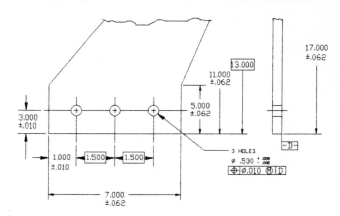

Fig. A-4

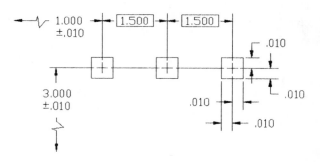

Fig. A-5

FEATURE TOLERANCE

These square tolerance zones provide the zone for each individual feature's circular positional tolerance zone. The center of the positional tolerance zone must lie on or within the boundaries of the square pattern plus/minus tolerance zone. The centers of the circular positional tolerance zone must also be at the specified basic dimension from each other. The two tolerance zones are illustrated in Fig. A-6.

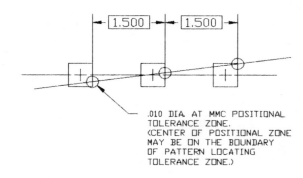

Fig. A-6

FEATURE LOCATION

The axis of each feature must lie somewhere within the circular tolerance zone. This means that the actual feature axis could lie outside the boundary of the square pattern positional tolerance zone. When the positional tolerance is applied on an MMC basis, any feature not at MMC is permitted an increase in positional tolerance, which, could allow the feature axis to be farther from the pattern tolerance zone boundary. The tolerance zone increase as the feature departs from MMC is determined the same for the 1973 ANSI

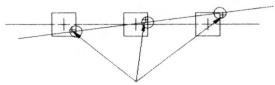

AXIS OF EACH FEATURE
(TOLERANCE ZONES AT .010 MMC)

Fig. A-7

standard as it is in ANSI Y14.5M-1982. Figure A-7 illustrates how the features may be positioned within the circular positional tolerance zone.

Symbols

GEOMETRIC DIMENSIONING AND TOLERANCING

C.'Bore or Spotface ——— ⌴

Depth or deep ————— ↧

Quantity ——————— ✕

C'Sink —————————— ⌄

Square ————————— □

Reference ——————— ⟨ ⟩

All Around ——————— ⌀⟋

Conical Taper ———— ⊳

Flat Taper Slope ——— ⊲

Radius ————————— R

Spherical Radius ———— SR

Spherical Diameter ——— SØ

GEOMETRIC DIMENSIONING AND TOLERANCING

Arc Length ——————— ⌒2.100

Maximum Material Condition MMC —— Ⓜ

Regardless of Feature Size ——— Ⓢ

Least Material Condition LMC —— Ⓛ

Projected Tolerance Zone ——— Ⓟ

Circularity ———————— ○

Straightness ———————— —

Cylindricity ——————— ⌭

Perpendicularity ————— ⊥

Flatness ————————— ▱

Profile of a surface ———— ⌒

Profile of a line ————— ⌒

GEOMETRIC DIMENSIONING AND TOLERANCING

Basic, or Exact, Dimension — [2.125]

Datum Feature Symbol — [-A-]

Feature Control Frame — [⊕|Ø.010 Ⓜ|A|B]

Datum Target — (A1)

Parallelism — ⫽

Angularity — ∠

Position — ⊕

Concentricity — ◎

Circular runout — ↗

Total runout — ⫮

Diametrical (Cylindrical)
Tol Zone or Feature — ⌀

Glossary

Actual size   The measured or produced size of a feature.

Allowance   The tolerance that permits two parts to be assembled with either a clearance or interference fit between parts.

Angularity   The condition of a plane or axis at a specified angle, other than 90°, from a datum plane or axis.

Axis   The theoretically centered line through the center of a cylindrical feature.

Barrelled   The condition of a cylindrical feature where it bulges in the center and tapers down to the ends.

Basic dimension (BSC)   A theoretically exact value specified on a drawing to describe the size, shape, or location of a feature. Variations from these dimensions are specified as feature tolerances, notes, or other dimensions.

Basic size   The size specified for a feature in which the size tolerances apply.

Bilateral tolerance   A tolerance specified by the designer that permits a dimension variation in two directions, e.g., plus or minus 0.5.

Boundary of perfect form   The condition (envelope) of true geometric form represented by the drawing.

Boss   A raised area on castings or forgings to allow more material for threading, stop, support area, or bearing area.

Callout   A specific note on the blueprint stating dimensions, tolerances, geometric controls, or feature specification.

Center plane   The plane or line located in the middle of a noncylindrical feature.

Characteristic   An integral part (a symbol) of the geometric dimensioning and tolerancing system or a feature of a part or an assembly.

Checking fixture   A go and no–go-type gage used to verify features.

Circular runout   The condition of a circular line element during a complete revolution where composite control in relation to the axis is measured at any cross section along the controlled surface.

Circularity   The surface condition of a feature (cone, sphere, cylinder) where all elements in a cross-sectional measurement must be an equal distance from the axis within the limits of a specified tolerance during one complete revolution.

Clearance   The maximum intentional difference between the size of mating parts.

Coaxiality   The condition where two or more axes are in alignment with each other.

Combined feature control frame   A feature control frame made up of two or more feature control frames each with a geometric characteristic symbol or a combination of feature control frame and datum feature symbols.

Composite feature control frame   A feature control frame made of two or more feature control frames with the same geometric characteristic symbol.

Concentricity   The condition where two or more features (holes or diameters) have a common axis.

Conical   Cone-shaped part of features.

Contour tolerancing   *See* Profile of a line or Profile of a surface

Control   A limitation specified by the designer for

various features. The limitation is specified as a size or geometric tolerance.

Coordinate   A set of numbers used to specify or determine the location from an X and Y axis of a point, line, curve, or plane.

Coordinate measuring machine (CMM)   An electronically controlled machine used to determine the location or condition of features in space. The data generated may be printed out with a printer option.

Coplanary   The condition of two or more surfaces in one plane.

Counterbore   A straight-sided recessed area around the end of a hole so a fastener is flush below the surface of a part.

Countersink   A taper sided recessed area around the end of a hole so a beveled fastener is flush below the surface of a part.

Cylindricity   A condition of a surface where all points on that surface are of equal distance from the axis of that surface.

Datum   A theoretically exact dimension origin. A datum may be a point, axis, line, or plane used for repeatable measurements.

Datum axis   The axis established by the intersection of the X and Y datum planes of cylindrical features, or the axis of a cylindrical feature established by the actual irregularities of the feature extremities. The axis is theoretical.

Datum feature   The actual part feature used to establish the datum.

Datum feature symbol   A box that contains the datum identifying letter with a dash on either side of the letter.

Datum line   The line derived from the actual counterpart of the datum feature. The line may be established by two planes, or the simulated center of cylinders or holes.

Datum of size   Any feature specified as a datum reference that is subject to size variation based on plus/minus tolerances such as widths or diameters.

Datum plane   A theoretically exact plane established with checking fixtures or gages when in contact with the counterpart of actual datum features.

Datum point   A theoretically exact point specified with a datum target of little or no size that has position on a surface for functional gaging purposes.

Datum reference   A feature specified on the drawing as a datum feature.

Datum reference frame   The three mutually perpendicular planes used to establish the theoretical datum planes for repeatable orientation from design to inspection.

Datum surface   A theoretical surface such as cylinders, slots, holes, edges, surfaces, etc. Used to establish repeatability.

Datum target   A specified line, point, or area identified with a datum target symbol to establish repeatability.

Diameter   Describes through the axis measurement of cylindrical features and tolerance zones.

Dimension   The numerical value specified for the size and or location of features.

Dimension line   The line drawn from the dimension value to the feature extension line.

Eccentric   A condition where two or more features do not have a common axis.

Error   A variation from a desired dimension or geometric form, location, or orientation that is unintentional. Errors are acceptable within the limits of the specified tolerances.

Extension line   The line used to extend the object line of parts.

Feature   The term given to any physical portion of a part, e.g., surface, hole, or pin.

Feature, functional   A feature controlled geometrically to meet specific design requirements. They are used to locate features from, and/or mate with other features that are interrelated in the overall design.

Feature of size   A feature dimensioned with a tolerance. The features may be cylindrical or noncylindrical.

Feature control frame   A rectangular box containing the specific instruction(s) for a feature or group of features. The rectangle contains the geometric characteristic symbols, tolerance, datum reference letters, and maybe the diameter and modified symbols.

Fit   The clearance or interference between two mating parts or fasteners.

Flatness   The condition of a surface having all elements in one plane.

Form   The desired finished shape for a given feature.

Form tolerance   A tolerance specified to allow a specified variation in a feature or surface from the desired perfect form.

Full indicator movement (FIM)   This term replaces older terms, Full Indicator Reading (FIR) and Total Indicator Reading (TIR). It is the full movement of an indicator needle while measuring a feature during a full rotation or complete travel along a feature.

Function   The consideration for the movement involved with assembled parts.

Geometric characteristics   The symbols specified for form orientation, profile, runout, and location tolerancing. These are the symbolic language of G.D.T.

Implied datum   An unspecified feature implied by dimensioning to be the origin for measurements.

Indicator   A precision measuring tool used when checking feature variations.

Knee   A piece of equipment used to rest parts against during machining and inspection operations.

Least material condition (LMC)   This term is a modifier that is specified with tolerances and datums. It means the condition in which a part or feature contains the least amount of material; for example, the smallest pin size or the largest hole size.

Leader   The line or arrow used to tie dimensions, symbols, etc., to part features.

Limit dimensioning   The specified maximum and minimum sizes of a feature.

Limits of size   The extreme minimum and maximum sizes permissible for a feature when considering the tolerances.

Location tolerance   A tolerance specified to allow a specified variation in the perfect location of a feature(s) as drawn on the drawing. The tolerance applies in relation to the datum features used to locate the controlled feature(s).

Maximum material condition (MMC)   This term is a modifier that is specified with tolerances and datums. It means the condition in which a part or feature contains the maximum amount of material; for example, the largest pin size or the smallest hole size.

Maximum dimension   The acceptable upper limit or the largest value specified for a feature.

Median plane   The center or middle plane of a noncylindrical feature.

Minimum dimension   The acceptable lower limit or the smallest value specified for a feature.

Minimum material condition   *See* Least material condition.

Modifier   In G.D.T. the term is used to modify or change specified feature tolerances. There are three modifiers: MMC, LMC, and RFS.

Multiple datum reference frames   Referred to when more than one datum reference is specified on the same part.

Nomional size   The stated designation that is used for the purpose of general identification of material, or the theoretical size of a feature, e.g., one inch bar or 1.000 plus or minus .010.

Optical comparator   An instrument for comparing a surface with an ideal surface or standard.

Origin   The location where dimensions and tolerances begin.

Orientation   The relationship of one feature as it relates to another specified as the datum.

Orientation tolerance   This tolerance is applicable to features that are related to another feature. The orientation controls are perpendicularity, angularity, and parallelism. Each of these must be related to a datum feature.

Pattern   A particular arrangement of features in a part.

Parallelism   The condition of a surface or axis at an equal distance from all points of a datum surface or axis.

Perpendicularity   A condition of a feature median plane, axis or surface which is 90° from a related datum axis or plane.

Pitch The distance from a thread point to the corresponding point of the next thread.

Plane A surface condition that is straight, flat, or level.

Position A location described by dimensions or the actual location of a feature.

Positional tolerance. *See* Location tolerance.

Primary datum A theoretical plane requiring three points of contact to establish that plane on an actual part. First in importance.

Profile of a line The condition where a permitted amount of profile variation, bilaterally or unilaterally along a line element, is specified.

Profile of a surface The condition where a permitted amount of profile variation, bilaterally or unilaterally over a surface, is specified.

Projected tolerance zone A zone specified for a given height above a hole in which a pin, cap screw, bolt, etc., is to be installed. The zone provides a cylindrical area in which the feature axis must lie. It controls the perpendicularity of the hole to prevent interference with the mating part.

Radius Describes a measurement from the axis to the surface of cylindrical features.

Reference dimension A dimension specified as an approximate dimension that is not toleranced.

Regardless of feature size (RFS) This term is a modifier that is specified with tolerances and datums. It means that, regardless of the actual feature or datum of size, the stated feature tolerance applies.

Relationship The consideration of an assembly in a static or fixed situation.

Roundness *See* Circularity.

Runout The composite variation from a desired surface during a complete rotation of the part around the axis.

Runout tolerance. A specified variation for an actual feature surface or line in relation to the axis during one complete revolution.

Secondary datum A theoretical plane requiring two points of contact to establish that plane on an actual part. Second in importance.

Size tolerance A tolerance specified to allow a feature to vary a specified amount.

Slope The inclination of a surface expressed as a ratio of the difference in heights at each end divided by the distance between those heights.

Specification A detailed description of a feature requirement such as size, shape, location, tolerance, etc.

Spherical A globular body, ball, having all points equal distance from a given center point.

Specified datum A datum feature identified with a datum feature symbol.

Spot face A machined surface of a specified size on a casting or forging.

Squareness *See* Perpendicularity.

Straightness The condition of a surface or axis where a single line element must be in a straight line.

Symmetry A condition in which a feature is equal on either side of a center line. Symmetry is no longer a current control in G.D.T.

Tertiary datum A theoretical plane requiring one point of contact to establish that plane on an actual part. Third in importance.

Three-plane concept The concept of three mutual planes exactly theoretically 90° and perpendicular to each other. Used for repeatable orientation.

Tolerance The total amount a specific dimension may vary. Between the maximum and minimum limits of size.

Tolerance zone A boundary established by a size or geometric tolerance in which the actual feature must be contained.

Total runout The condition of all surface elements during a complete revolution where composite control in relation to the axis is measured.

True position An exact (perfect) location described by basic dimensions in relationship to a datum or other feature. The location specified on the drawing.

Unilateral tolerance A tolerance that permits variation in one direction from the specified dimension, e.g., 1.000 plus .000, minus .010.

Virtual condition The condition of a feature where the collective effect of size and form error establish the feature size required for determining fit between parts.

Weldment A unit or assembly formed by the welding together of pieces.

Wristed The condition of a cylindrical feature where it narrows toward the center from each end.

Answers to Questions

Chapter One

Answers: 1. communication; 2. symbols; 3. American National Standards Institute (ANSI); 4. clarity; 5. replace; 6. total; 7. form; 8. size, location; 9. function, relationship; 10. tolerances, interchangeability; 11. tolerance; 12. plus/minus

Chapter Two

Answers: 1. 17; 2. 2; 3. 5; 4. 10; 5. 19; 6. 6; 7. 5; 8. 1; 9. 22; 10. 20; 11. 7; 12. 3 & 4; 13. 8; 14. 21; 15. 11; 16. 14; 17. 23; 18. 12; 19. 15; 20. 16; 21. 9; 22. 6; 23. 17

Chapter Three

Answers: 1. specified; 2. dimension, orientation; 3. surface (planes), axis, centerlines, edges (lines), points, areas; 4 no; 5. three; 6. 90; 7. circle; 8. size; 9. lines, points, areas; 10. tolerance; 11. simulated; 12. modifiers; 13. convey; 14. rock; 15. flat.

Chapter Four

Answers: 1. control; 2. combined; 3. surface; 4. geometric characteristic; 5. tolerance; 6. left, right; 7. primary; 8. line; 9. compartments; 10. no.

Chapter Five

Answers: 1. size, form; 2. individual; 3. modifiers; 4. MMC; 5. zero; 6. RFS; 7. pitch; 8. beneath; 9. foundation; 10. not

Chapter Six

Answers: 1. directions; 2. radially; 3. element; 4. boundaries; 5. entire; 6. error; 7. tolerance; 8. diameter; 9. attachment; 10. implied; 11. assumed; 12. size, geometric; 13. axis; 14. distance; 15. width; 16. RFS; 17. bilateral; 18. line; 19. profile; 20. revolution

Chapter Seven

Answers: 1. collective; 2. geometric; 3. exceed; 4. gages; 5. acceptable; 6. mating; 7. opposite; 8. interchangeable; 9. five; 10. modifier

Chapter Eight

Answers: 1. position, concentricity; 2. variaton; 3. RFS; 4. cylindrical, noncylindrical; 5. basic; 6. specified; 7. increases; 8. shift/rotation; 9. axis; 10. length, depth; 11. virtual condition; 12. MMC; 13. MMC; 14. size; 15. noncumulative

Chapter Nine

Answers: 1. D = .005; E = .040; 2. D = .003; E = .003; 3. RFS; 4. Yes; Rule 3; 5. 2.025 × 1.025 × 4.00; 6. .005; 7. .002; 8. Rule 1—the size tolerance; 9. .475 × 1.975; 10. No; 11. datum feature; 12. Noncylindrical; the diameter symbol is not specified preceding the .015 tolerance; 13. Width; 14. Yes; parallelism; 15. surface; 16. Total; 17. FIM; 18. total; 19. One; 20. axis; 21. irrelevant; 22. composite; 23. diameter; 24. surface; 25. line; 26. bilateral; 27. side; 28. surface; 29. .010; 30. irregular; coplanar; 31. .040; .030; 32. 1.462; 33. Yes; 34. attitude; perdendicularity; 35. The basic dimensions from datum surfaces "B" and "C"; 36. Yes; 37. .528 diameter; .538 diameter; 38. .498 diameter; 39. pattern; 40. .010; 41. may; 42. must; 43. .010; 44. must; 45. control; attitude; 46. pass; 47. .010; .020; 48. .313; 49. It is a width; 50. Yes; 51. Because it is a feature of size; 52. No. Basic dimensions do not have tolerances. The spacing is permitted to vary based on each feature's tolerance; 53. composite; 54. pattern; 55. .020; 56. each; 57. Because they are not features of size; they are all surfaces.

Index

guy